MINERALS AND TI

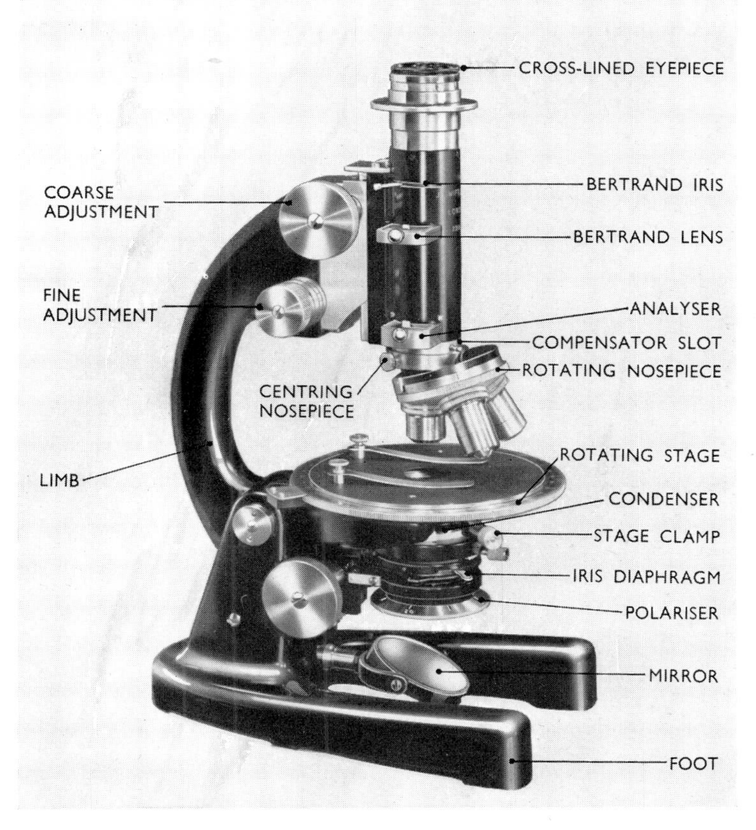

By courtesy of James Swift & Co., Camberwell Road, London, S.E.5.

Fig. 1. Petrological microscope.

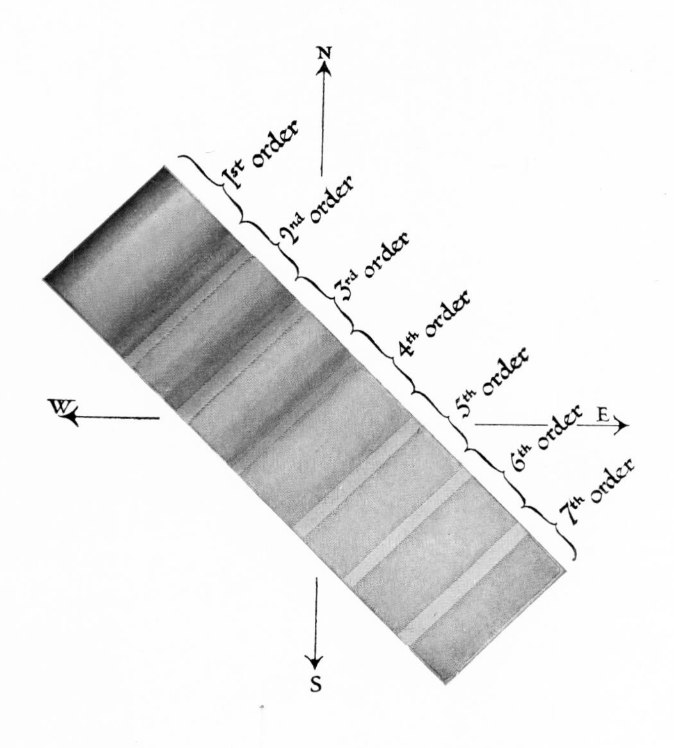

Newton's Scale

The colours are seen when a wedge of quartz or gypsum is placed in an oblique position between crossed nicols.

The points of the compass illustrate the scheme adopted in order to indicate directions in the field of the microscope, it being assumed that the observer always faces north.

H. G. SMITH

MINERALS and the MICROSCOPE

Completely Revised
by

M. K. WELLS
M.SC., PH.D., D.SC.
University College, London

London
Thomas Murby & Co

FOURTH EDITION 1956
REPRINTED 1957, 1960, 1964, 1968, 1971

ISBN 0 04 541002 3

George Allen & Unwin Ltd.
40 *Museum Street, London, W.C.*1
are now the proprietors of
Thomas Murby & Co.

PRINTED IN GREAT BRITAIN
BY OFFSET
UNWIN BROTHERS LIMITED
WOKING AND LONDON

PREFACE TO FOURTH EDITION

The late H. G. Smith succeeded to a notable degree in providing several generations of students of Geology with a concise and understandable introduction to the study of minerals in thin sections under the microscope. The present writer has tried to keep the subject matter within the bounds of previous editions, but some expansion has been necessary. Much of the book has been re-written, partly because of changes in the design of petrological microscopes, partly as a consequence of the growth of knowledge on the subject.

Formerly the polarising apparatus below the stage of the microscope was easily removed from the light-path, and it was therefore appropriate in earlier editions to include a chapter on the appearance of thin sections of minerals in ordinary (non-polarised) light. In most modern instruments, however, the polariser is used as a fixture: polarised light is invariably used in the study of thin rock- and mineral-sections, and such a chapter becomes unnecessary.

The writer believes that a good working knowledge of all the apparatus used is essential in the training of the geological student. Therefore the one-page description of the petrological microscope of former editions has been expanded, so that the beginner may make the most efficient use of the instrument (which is a precision instrument) at his disposal.

For similar reasons the specialised techniques used in the study of sections of minerals are rather more fully described and explained than formerly.

The most important development in mineralogy within recent times has been the application of X-ray techniques to the study of the internal structure and detailed composition of minerals. As a result it is now practicable (and desirable) to discuss the composition of, say, the all-important rock-forming silicates in terms which can be understood by everybody. A section dealing with this aspect of mineralogy has therefore been included in the new edition.

In the description of each mineral a short paragraph has been devoted to modes of occurrence, as essential 'background' knowledge: the associations of minerals are as important as their specific properties.

The thirty-page introduction to petrology found in previous editions has been deleted, as it is considered out of context in a book with the title, *Minerals and the Microscope*.

The writer is grateful to his father, Dr. A. K. Wells, for criticism and help in the preparation of the new edition, and also to his colleagues at University College, London, for assistance rendered.

August, 1955. MAURICE K. WELLS.

CONTENTS

PLATES

CHAPTER 1

Introduction

Some rock specimens may appear at first glance to be somewhat dull and uninteresting objects. It may be possible with the naked eye to see the outlines of the component mineral grains which make up a rock, and in certain fortunate cases individual minerals may be identified by their characteristic shapes or colour or lustre. On the whole, however, many rock specimens remain uninformative and probably rather uninteresting to anyone who has not specialised in their study.

With the aid of a petrological microscope, these same rocks, suitably sectioned, become objects of infinite fascination. Most of the rock-forming minerals become transparent and lightly coloured, though often dark and sombre coloured, or even apparently opaque in the mass. Each mineral is found to produce its own distinctive effects on the light that is transmitted through it. Minerals may be identified by their characteristic shapes and cleavages; by their colour (which is generally a much more reliable criterion than in hand specimens, where impurities in the mineral may completely mask the true colour); by the boldness with which the outlines of some minerals stand out in contrast with their fellows; and by inclusions and alteration-products. A petrological microscope is, moreover, much more than an efficient magnifier in that it has a device, appropriately termed an analyser, which can be placed above the thin-section to transform normally colourless minerals into objects of unexpected and often rich colouring. The proper interpretation of these so-called interference colours may be quite a difficult matter. It is obvious, however, that the production of these colours provides a whole new set of criteria by which minerals may be distinguished.

It is the object of the writer to describe and explain the phenomena produced by 'minerals and the microscope' in as simple a way as possible. Some features relating to the construction and use of the microscope are described in the first chapter. In previous editions this section was restricted to a brief appendix; but the present writer's experience is that many of the difficulties encountered by students in the microscopic study of minerals are due more to lack of familiarity with the instrument than lack of knowledge of principles of crystal optics. In the next three chapters those principles are described. The various phenomena that will be encountered by the student are described in nearly every case before the explanations are given: effects are described before causes. In this way the book can be used as a practical aid. Ideally the student should carry out the various tests mentioned in the text and see the effects for himself. It is not intended that the book should be read as a theoretical text on crystal optics. There are already several excellent books on this subject, including those listed in the bibliography on p. 143.

The arrangement of the first four chapters has been chosen in accordance with the increasingly specialised use of the microscope and its accessories. All petrological microscopes have a so-called polariser through which the light must pass before reaching the thin section. As a result the light is restricted in its direction of 'vibration' and is said to be plane polarised (see p. 16). The simplest normal method of viewing minerals with a petrological microscope is, therefore, with plane polarised light (Chapter 4). The effects of the polarisation are not generally obvious, and for many minerals their appearance in ordinary light and polarised light is identical. It is only when the polariser is used in conjunction with the analyser that the most striking effects are produced (Chapter 5). Lastly, there are certain more specialised tests that can be carried out with convergent light as described in Chapter 6. The microscope has to be adapted with the aid of extra lenses so that the light entering the mineral is no longer in the form of a parallel beam, but forms a cone of rays. The image seen is then not a magnification of the original object; but a symmetrical 'interference figure' or optic picture.

A petrological microscope, therefore, enables a considerable number of tests to be applied to transparent minerals. The tests themselves make a fascinating study, and because of their diversity they make it possible for most minerals to be identified with comparative ease.

CHAPTER 2

The Petrological Microscope

In the following description stress is laid upon hints of a practical kind that will help a beginner to make the best use of a petrological microscope. No attempt is made to give technical details, or descriptions of the principles of optics that govern the design of a good microscope (fig. 1).

There are three main sets of components in a petrological microscope:

1. *The stand,* consisting of a heavy metal base or foot, to which is attached the body tube and the rotating stage.
2. *The optical system* consisting of a tilting mirror, various substage lenses and possibly a converger lens, the objective at the lower end of the body tube and the eye-piece or ocular at the top.
3. *Devices for producing plane-polarised light.* These are peculiar to the petrological microscope and consist of a lower **polariser** mounted below the microscope stage, and an **analyser** of similar design mounted on the body tube above the objective. The operation of these two units will only be appreciated fully when the reader has obtained a considerable knowledge of the optical properties that are peculiar to minerals, as described in the early chapters of this book. The polariser is generally designed as a fixture in the normal light-path of the microscope, while the analyser is mounted on a sliding or hinged unit so that it can be removed from the light-path in a moment. If care is taken to ensure that the analyser is so removed, then it will be found that the polariser has no obvious effect on the normal operation of the instrument as a high-power magnifier.

THE MICROSCOPE STAND

The stand comprises nearly all the mechanical parts of the microscope. It is not generally realised by the student how completely the efficient working of a microscope depends upon the exact fit and proper lubrication of all the moving parts. These include the joint by which the tube and stage assembly can be pivoted to suit the eye-level of each individual worker. Generally the desired tilt is maintained by some degree of friction in the joint, and if this becomes too slack due to wear, so that the tube tends to tilt too easily and of its own volition, it is possible to tighten the mechanism by an adjusting screw. On more expensive instruments there may be a locking lever, in which case it is important to ensure that the lever is slackened before tilting the tube, and locked afterwards.

The rotating stage is a part of the microscope which requires expert lubrication. It should be capable of rotation with application of very little pressure. In some microscopes a screw-operated clamp is provided to fix the stage in any desired position. These clamps can easily be damaged by over-tightening. Stage clips are provided to hold a thin-section or mineral mount in place. These clips may bear down heavily on the glass slide, and because of their sharp edges it is easy for them to damage the cover-slip of a section. For this reason, when it is necessary to move a section about on the stage in order to select a particular grain or rock-texture for viewing, it is better to hold the section lightly between finger and thumb and to apply the clips only after the section is in its desired position.

The tube of the microscope is attached to the remainder of the stand by means of a focusing rack. Focusing is achieved by adjustment of coarse and fine focusing knobs. When a microscope is provided with a double or triple nose-piece, the focal lengths of the various objectives are such that they can be interchanged without re-focusing except perhaps for a touch on the fine adjustment when a high-power objective is being used.

It sometimes happens that the focusing mechanism becomes slackened through use so that the tube may shift slightly under its own weight. This can generally be remedied by the turn of a screw; but as with every other mechanical adjustment of a microscope, this is a matter for expert attention. It should not be necessary to say that a petrological microscope is a precision instrument which should always be handled with great care. In particular, none of the moving parts should ever be forced.

THE NORMAL OPTICAL SYSTEM: THE OBJECTIVE AND OCULAR

Every kind of microscope has a normal optical system consisting of a reversible plane or concave mirror, two or more interchangeable objectives and possibly a choice of oculars or 'eye-pieces'. The different **objectives** are required for varying magnification. The main point to remember in this connection is the fact that a low-power objective is easier and more restful to use than one of high power. There are two reasons for this. In the first place, the focusing is much less critical with low-power magnification. With a really high magnification the depth of focus is not sufficient to bring the full thickness of the rock slice (0·03 mm.) into view at one time. Secondly, a high-power objective has a very small aperture, so that a restricted quantity of light has to illuminate the whole of the magnified image. This gives a low intensity of illumination unless an extra-strong light source is used. Apart from these technical advantages that are gained by the use of fairly low-power objectives, there are also obvious advantages in being able to see a large and representative part of a rock-section at a single glance. *It is a useful general rule to start any rock or mineral study with the lowest-power objective available, and to increase the magnification later when it becomes necessary to examine increasingly minute details of interest.*

Objectives may have their magnification engraved on them, or alternatively the focal length may be quoted. Two commonly-used focal lengths are those of 1 inch and $\frac{1}{4}$ inch which correspond to magnifications of about $\times 5$ and $\times 30$ respectively. A rough idea of the power of an objective can be gained from the diameters of the lenses: those with the smallest diameter correspond to the highest magnification.

When the microscope has a double or triple nosepiece it should only be necessary to make a single focusing adjustment. One practical point to remember about focusing with a high-power objective is the very small clearance between the cover-glass of the section and the objective. Forgetfulness on this point has caused damage to many thin-sections and no doubt to objectives as well!

Total magnification produced by the microscope is obtained by multiplying the magnification factor of the objective by that of the **eye-piece** or **ocular.** There are two special features to note about the latter. Firstly, two fine cross-wires or cross-hairs are included in the assembly. These act as reference lines for measurement of directional properties of minerals, and they are used in conjunction with the graduated rotating stage as described below. It is, therefore, very important for the cross-wires to be in their correct position in relation to the rest of the microscope. This is the reason for a second special feature of the ocular of a petrological microscope: the presence of a small projection on the latter which fits neatly into a corresponding slot cut in the top of the microscope tube, thus holding the ocular in its correct position.

When a thin section is correctly in focus it should be possible to see the magnified image of the cross-wires in what appears to be the same plane as the object-image. The field of view is then divided into four quadrants by the cross-wires which are commonly referred to as extending in the east - west and north - south directions. Some oculars are fitted with independent focusing for the cross-wires.

CENTERING THE OBJECTIVES

Precise alignment of all parts of the optical system relative to the axis of the microscope is an obvious necessity. It is particularly vital when the microscope has a rotating stage.

Any object lying 'at the intersection of the cross-wires' should remain in that position when the stage is rotated. If it moves away from the centre, the position of the objective needs adjustment. In most microscopes two screws may be found projecting from the tube just above the objective or nosepiece (fig. 1). These form part of the mounting of the objectives, and by tightening or slackening the appropriate screws the latter can be centred (fig. 2). Before undertaking this centering process it is best to ensure firstly that the nosepiece has been 'clicked' into position, and secondly that the internal springs of the centering mechanism have not become jammed. Either effect would normally cause the image to follow a wildly eccentric path during stage rotation.

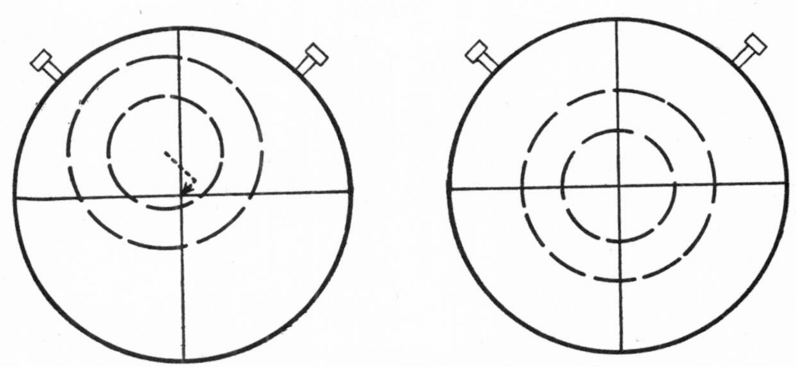

FIG. 2. *Centering an objective.*

If centering is necessary with a multiple nosepiece instrument, it should always be adjusted for the highest power objective. Obviously in the small but highly magnified field of the latter, a slight maladjustment, which would only be a nuisance with lower magnification, could carry the object right outside the field of view.

ILLUMINATION

One of the problems that confronts the student learning to use a petrological microscope is that of getting adequate and uniform illumination over the whole field of view. All too often the field is dark or only illuminated in one patch, sometimes with the appearance of a reduced image of a window or the trade-name of an electric-light bulb! If the following points are checked, it should be possible to get reasonable illumination.

1. Align the microscope with the lamp.
2. Ensure an adequate spreading of the light over the whole field by having the plane side of the mirror uppermost; by varying the spacing between lamp and microscope; and if necessary by using a ground glass filter to diffuse the light evenly and to remove the unwanted images mentioned above. This glass may be placed in front of the lamp or in a holder beneath the microscope stage.
3. Check the position of the substage converger. In some microscopes this is permanently in position; in others it should be removed from the light path.
4. Withdraw the analyser. When in position this automatically blacks out the light.
5. If a Bertrand lens (see p. 62) is fitted, this also should be removed from the light path.
6. Open the substage diaphragm (when present) to a reasonable degree.
7. Always commence work with a low-power objective which passes much more light than one of high power.

The polariser and analyser, which form such an important part of a petrological microscope, are described in Chapter 3. In order to understand their functions it is necessary to know something about the principles that govern the vibration of light in crystalline substances. The polariser restricts the light that passes through it to a single plane of vibration. This has very little effect on the appearance of most minerals seen in thin-section. Therefore, although the polariser is a permanent fixture beneath the stage of most petrological microscopes, and all the light has to pass through it, its presence can be ignored so far as the adjustment of the microscope is concerned. The effects of the analyser (essentially a second polariser, but mounted in the body-tube with its plane of vibration at right angles to that of the polariser) are much more striking. These cannot be ignored, and during the setting up of the microscope the analyser has to be withdrawn from the light path.

CHAPTER 3

Some Principles of Refraction and Double Refraction in Crystals

THE NATURE OF LIGHT

It is only necessary to have a very simplified idea of the nature of light in order to understand most of the optical effects produced in crystalline and transparent minerals.

As light may be transmitted through a vacuum, it is necessary to assume the existence of a substance pervading all space, by means of which the transmission is effected. This substance is known as the ether. It is quite certain that light is transmitted as a consequence of vibrations which take place at right angles to the direction in which light is travelling, and the theory formerly accepted was that ether 'particles' themselves performed these vibrations to and fro without undergoing any movement along the line of propagation of the ray. The theory has now been considerably modified, and a theory of oscillation, not of the particles themselves, but of their electrical condition, is now held to account for the facts more satisfactorily. But these oscillations or vibrations certainly do take place, and they take place at right angles to the direction of propagation of the light. A ray of ordinary light performs its vibrations in all directions possible subject to this limitation, that is in all directions perpendicular to the direction of transmission. Light vibrations take place in exactly the same way in all gases, in liquids and in glass.

VELOCITY OF LIGHT AND REFRACTION

Every substance, of course, affects the velocity of the light vibrations to some extent, and a comparison of the speed of light in a particular medium with its speed in air, provides one of the valuable and measurable properties of the medium known as its **refractive index.** The refraction or bending of a beam of light (that is, of a 'bundle' of parallel rays) follows inevitably from any change in velocity that light suffers in its passage from one medium to another. This fact is illustrated in fig. 3. For bending to occur, the

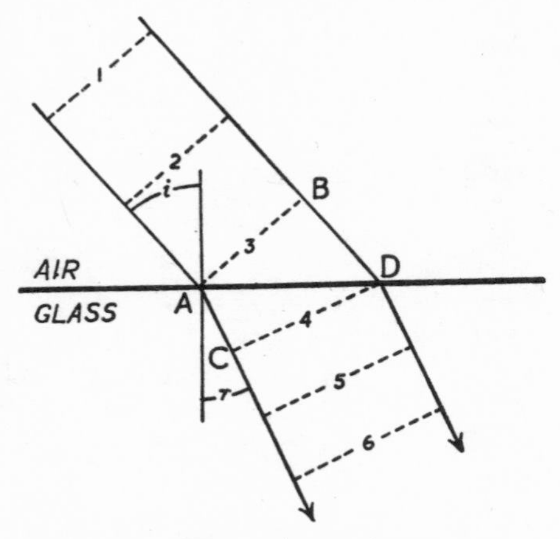

FIG. 3. *Refraction. For explanation see text.*

incoming, or incident rays must be oblique to the surface separating the two media. This obliquity is measured from the normal (or direction perpendicular) to the surface, and is known as the *angle of incidence, i.* The corresponding *angle of refraction* is *r.* At a given instant in time the incident wave-motions would reach the positions shown by A and B, the line AB representing the so-called wave-front which is perpendicular to the incident rays. While the fast-travelling light in air continues to advance from B to D, the slower travelling light in the glass covers a shorter distance,

AC. The wave front in the glass is therefore pulled round to a new direction and the rays are refracted. The ratio $\frac{BD}{AC}$ is the ratio:

$$\frac{\text{velocity of light in air}}{\text{velocity of light in glass}} = \text{refractive index, } n\text{, of the glass.}$$

For those who wish to express the relationship mathematically, it is easy to show that:

$$\text{refractive index, } n = \frac{\sin i}{\sin r}$$

This is an important relationship which is the basis of all practical refractive index measurements. It expresses the refractive index in terms of measurable angles rather than velocities, which of course, for light, with a speed of approximately 186,000 miles per second in air, are incapable of direct measurement.

One further aspect of refraction needs mention at this point. Light passing from a medium of higher to one of lower refractive index is bent away from the normal. As the angle of incidence is increased, the angle of refraction increases to a maximum of 90°. When this happens the angle of incidence is known as the **critical angle,** and once this angle has been exceeded, refraction ceases and reflection takes its place. Since the reflected rays return to the medium of the incident rays, the phenomenon is known as **total internal reflection.** Measurements of critical angles are used for determination of refractive indices; and use is made of total internal reflection in the Nicol prism described below.

ISOTROPIC AND ANISOTROPIC SUBSTANCES

Light is free to vibrate in all possible directions in air (or any other gases), in liquids and in glasses. The velocity of the light depends solely upon the refractive index of each medium, and is independent of direction. This uniformity of behaviour is due to the random distribution of the atoms forming each of the substances mentioned. Every substance of this kind which is singly refracting (*i.e.* has a single refractive index) is said to be **isotropic.**

In crystals, however, the situation is generally different. The atoms are arranged in orderly and geometrical three-dimensional patterns. They tend to exert an influence upon the vibrations of any light that passes through a crystal space-lattice. Instead of being free to vibrate in all possible directions, the light becomes regimented and forced to vibrate in certain planes only. An analogy is provided by a group of soldiers drawn up on parade. These correspond to the atoms in the space lattice. A light ray penetrating the lattice would be equivalent to a person trying to cross the parade ground. He would find two easy ways to cross: parallel with either the ranks or the files of soldiers. These directions would correspond to the planes of easiest light vibration in the lattice.

All crystalline substances which limit the freedom of light vibrations in this way are **anisotropic.** The light which is forced to vibrate in certain planes is said to be **polarised.**

It will be obvious from the description given above that the space-lattice of a crystal has a controlling influence upon the behaviour of light transmitted through it, and as might be expected, the symmetry of the crystal is reflected in the symmetry of its optical properties. This fact is brought out in Fig. 4 in which shapes representative of the seven crystal systems are drawn and the behaviour of light rays entering the crystals in various directions are indicated. It will be noticed that two symbols are used in the diagram. The asterisk symbol indicates that light entering the crystals in the direction shown behaves as though the crystal were isotropic: that is to say, the light vibrates with equal ease in all planes. This applies to *all* the directions in a Cubic mineral, and to the vertical axes of Tetragonal, Hexagonal and Trigonal minerals. For all other directions the symbols used in the diagram consist of crosses formed by two lines which are mutually at right angles to one another. These lines represent two planes of polarised light. It is a fact which we shall not attempt to explain here, that anisotropic substances always split up the light into two rays which vibrate in planes exactly at right angles to one another (see p. 21).

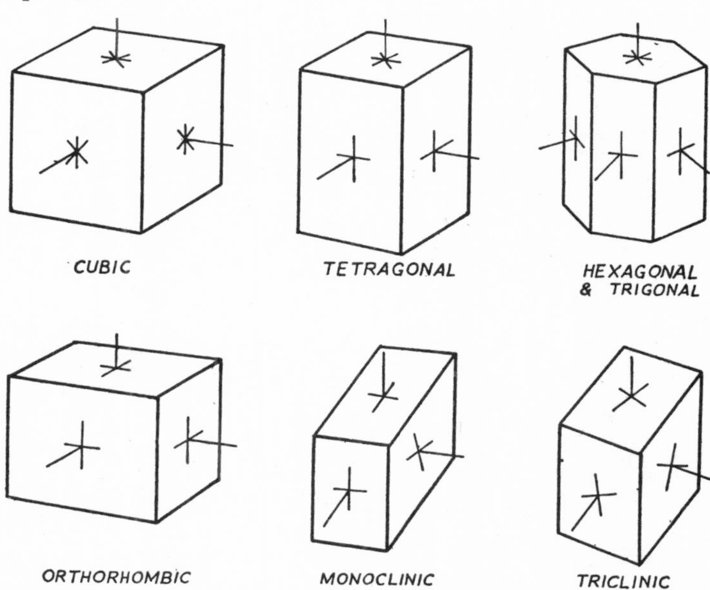

FIG. 4. *Relationship between crystallographic and optical symmetry shown diagrammatically. Light rays are pictured entering crystals from various directions (actually along crystal axes for convenience). The asterisk symbol indicates isotropic conditions, and the two vibrations of anisotropic conditions are shown as two lines intersecting at right angles.*

Fig. 4 is fairly self-explanatory and from it one may deduce a general rule relating to the behaviour of light entering a crystal (or, of course, it may be a thin slice cut from a crystal). If the direction of entry is along a crystallographic axis of more than twofold symmetry the light behaves as though the crystal were isotropic. In all other directions the crystal is anisotropic. It is interesting to find that a highly symmetrical arrangement of atoms, such as exists in Cubic minerals, has the same effect on light as the completely random distribution of atoms in the other isotropic substances mentioned above, namely gases, liquids and glasses.

The relationship between crystallographic and optical properties may be summarised as follows:

Crystal system	*Optical character*
Cubic	Entirely isotropic.
Tetragonal Hexagonal Trigonal	Single isotropic direction for rays travelling along vertical crystal axes. Remaining directions anisotropic.
Orthorhombic Monoclinic Triclinic	Anisotropic generally for rays travelling in all directions.

An excellent demonstration of the optical characters of an anisotropic mineral is provided by calcite. With the clear variety of calcite known as Iceland Spar, one can actually see the splitting of the light into two separate rays. Calcite has an atomic structure which provides the light with two paths, one of easy vibration where the velocity of the ray is great, and one difficult path where the velocity is much reduced. The two rays produced by an anisotropic substance always have different velocities and hence by definition they must have different refractive indices. They are therefore said to be doubly refracted. The phenomenon of **double refraction** is exhibited by calcite to a very marked degree.

THE DOUBLE REFRACTION OF CALCITE

The reader will find it most instructive to examine the behaviour of light as it passes through a fragment of calcite. Ideally a glass-clear rhombohedral cleavage fragment should be available. Light entering such a cleavage fragment perpendicular to one of the cleavage planes is split into two rays, and, in consequence, any object viewed along such a perpendicular is doubled: two images reach the eye. The most satisfactory demonstration of this fact is obtained by resting the fragment on a sheet of white paper marked with a black dot. Seen from above, such a spot is doubled (fig. 5). When the fragment is rotated, it is seen that one of

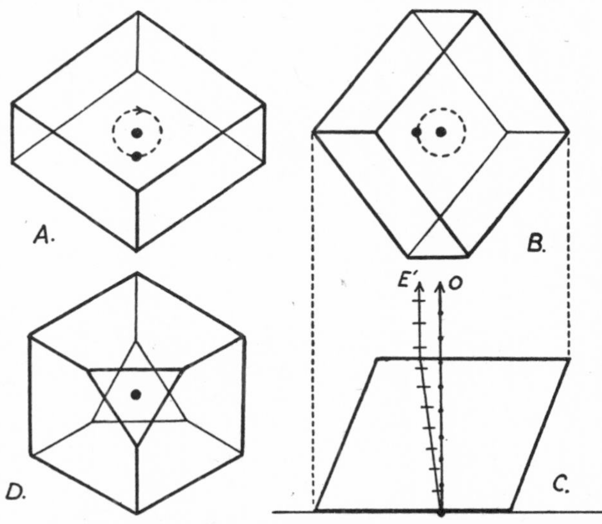

FIG. 5. *Double refraction of calcite. A and B show the double image seen through a cleavage rhombohedron. The formation of ordinary (O) and extraordinary (E′) rays is shown in the section of the rhombohedron, C. Single refraction seen along the vertical crystallographic axis of calcite is indicated by the single image shown in D.*

the images remains stationary, while the path of the other is the circumference of a circle with the stationary image as centre. The stationary image behaves just as if the crystal were a slab of glass; it is therefore said to be produced by the **ordinary ray.** The ray producing the movable image, on the other hand, does not obey the ordinary laws of refraction. To anyone who has studied only singly refracting (isotropic) substances, its behaviour appears to be quite extraordinary, and it is appropriately called the **extraordinary ray.**

By looking at the surface obliquely, one can see that the ordinary image appears to occupy a higher position inside the fragment than does the extraordinary image. In other words, the optical 'density' of the crystal for the ordinary ray is greater than for the extraordinary ray; or, the velocity of the extraordinary ray is greater than that of the ordinary ray. The fact that two such rays travelling in a doubly refracting crystal have different velocities, is extremely important to the student who wishes to understand what follows, and this experimental demonstration of the fact should help materially to fix the ideas.

Two more points of interest can be demonstrated with a little extra apparatus. It is possible to produce two surfaces parallel to the basal pinacoid by grinding and polishing on the two opposite obtuse corners of a cleavage rhombohedron of calcite. This is a job for a specialist since calcite is so soft and so easily broken along the cleavages. If one of these artificial basal pinacoids is now placed over the ink dot, only one image will be seen. This shows that light transmitted along the direction of the vertical axis of calcite is only singly refracted and that calcite behaves as an isotropic substance in this direction. Calcite is a Trigonal mineral and therefore its optical properties fall within the second category listed on p. 18.

The last test involves the use of a polariser in conjunction with a cleavage rhombohedron of calcite. The function of a polariser is described below. Briefly it cuts out one of the two rays of plane polarised light. If the two images seen through the calcite are viewed through a polariser, and the latter is slowly rotated, first one dot and then the other will disappear. If a record can be kept of the rotation it will be found that the disappearance of one dot occurs every 180°, and that 90° separates the elimination of alternate dots.

This may be regarded as a demonstration of the fact that the two rays forming the two images are actually polarised at right angles to one another.

Calcite can therefore be used to demonstrate the behaviour of light transmitted through an anisotropic crystalline substance, as follows:

1. Two rays are generally produced.
2. Double refraction occurs so that the rays are bent to slightly different degrees (much more markedly in calcite than in most minerals).
3. The rays differ in velocity.
4. They are polarised with vibrations that are at right angles to one another, and to the direction of travel of the rays.
5. Calcite is one of the minerals that has a single isotropic direction.

FUNCTION OF A POLARISER

The effects of double refraction, and the production of two distinct rays each with its own plane of vibration, are self-evident in a cleavage fragment of calcite. Every anisotropic mineral splits the light into two separate rays; but because the double refraction is generally much less than that of calcite, the rays are not usually visibly separated. The eye sees the light transmitted by both rays simultaneously, and cannot distinguish between them. Any device that can isolate one or other vibration is, therefore, a valuable adjunct to a mineralogist. In a petrological microscope this effect is produced by a polariser.

A polariser makes use of the property of double refraction, but is so designed that only one of the two rays produced is eventually transmitted. The light that emerges is therefore confined to a single vibration direction in one plane and is said to be *plane polarised*.

There are now two widely-used types of polariser. The earlier kind is made of clear calcite (Iceland Spar), and is named the Nicol prism after its inventor. The second type of polariser is formed of a synthetic crystalline substance known as 'Polaroid'. This material is widely known since it is used in anti-glare spectacles by motorists and others. The end product of plane polarised light is the same with either a nicol prism or polaroid disc; but the two methods of effecting the polarisation are quite distinct, as described below.

THE NICOL PRISM

In this, use is made of the exceptionally wide divergence of the two rays produced by the strong double refraction of glass-clear calcite. Construction of one type of Nicol prism (or 'nicol' as it is often called) is shown in fig. 6. Two carefully cut and polished triangular prisms of calcite are

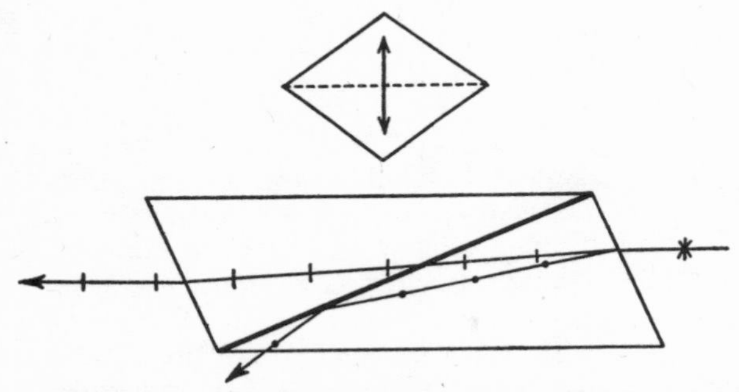

FIG. 6. *Operation of Nicol prism polariser.* (*a*) *Transverse section.* (*b*) *Plan view showing vibration direction of polarised light.*

cemented together by a film of Canada balsam. The angles of the prisms are of vital importance. They ensure that the incident light which enters the nicol is split into two rays which are as widely divergent as possible, and further that the one refracted ray meets the film of Canada balsam at such an oblique angle that it is totally internally reflected back into the lower calcite prism. Generally the sides of the nicol are surrounded by a black matt-surfaced material which serves to absorb this unwanted ray completely. The second ray (which is, of course, plane polarised by its passage through the atomic lattice of the calcite) is bent somewhat in passing through the Canada balsam film, but it resumes its initial course in the upper half of the nicol.

In the description of the double image which is seen through a cleavage rhombohedron of calcite given on p. 18, it was shown that the velocity of the extraordinary ray is faster than that of the ordinary ray. Because of this the ordinary ray is refracted more than the extraordinary one (see p. 14 for an explanation). Therefore in the Nicol prism it is the ordinary ray which is eliminated and the extraordinary one which forms the plane polarised transmitted light.

POLAROID

This is a crystalline substance which is almost colourless when mounted in the thin discs used for polarisation. The crystal orientation is uniform over the whole area of each disc. Light passing through it is forced to vibrate in two

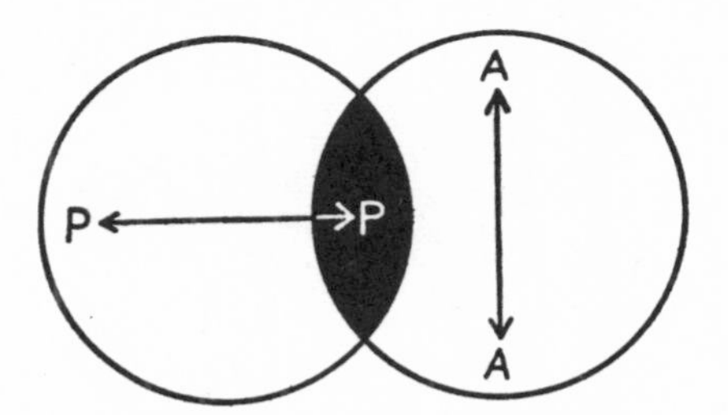

Fig. 7. *Diagram to show effect of crossing two polaroid discs. PP and AA represent vibration directions of transmitted rays of polariser and analyser respectively.*

planes as in any other anisotropic substance. One of the two rays is then completely absorbed by the polaroid, while the other passes through with negligible absorption. In effect the polaroid is opaque to one ray and almost perfectly transparent to the other (fig. 7). Certain minerals, such as biotite and tourmaline, have a similar property of differential absorption (= pleochroism, p. 33); but no natural substance has it to the extraordinary degree exhibited by polaroid. Because of this property, the latter forms a ready-made source of single-ray plane-polarised light.

CHAPTER 4

The Examination of Minerals with Ordinary and Plane Polarised Light

---

In previous editions of this book, separate chapters were devoted to the study of minerals in ordinary light and in plane polarised light. This was a logical arrangement which assumed that the polariser of the microscope could easily be removed from the light path as a matter of routine. In earlier models of petrological microscopes this was often the case; but nowadays the polariser is generally meant to be a permanent fixture. In other words, examination of minerals with ordinary transmitted light is now omitted from the routine. Only plane-polarised light is used. The appearance of most minerals in thin-section is the same under these conditions as it is with ordinary light. The main exceptions are provided by some coloured minerals which are pleochroic: that is, they change in colour when the stage is turned and when plane polarised light is used, as described on p. 33.

SHAPE

Rocks consist of aggregates of crystals or mineral grains which generally lie in all possible orientations. Sometimes, as in rapidly chilled volcanic rocks, well-shaped (euhedral or idiomorphic) crystals are embedded in a very finely crystalline or even glassy matrix. More generally the various minerals have interfered with each other's growth so that they form granular aggregates in which very few grains have anything approaching their true crystal shape. The grains are then said to be anhedral or xenomorphic.

In sedimentary rocks such as sandstones, crystals of quartz, etc., may have lost such perfection of crystal form as they once possessed by mechanical rounding and attrition. In such rocks there is often a secondary 'cement' of finely crystalline quartz, of calcite or of various iron compounds. The individual mineral grains forming the cement matrix will have their shapes controlled largely by the presence of the pre-existing detrital grains.

We see from the above that the shape of a mineral grain is largely a matter of its history and origin, and that it may vary from that of a perfectly developed crystal form to a completely irregular shape.

Because of the variable orientation of the minerals in a rock-section it is essential to acquire the habit of thinking in terms of three dimensions. The simplest of crystal forms, for example the cube shown in fig. 8, can yield a considerable diversity of shapes in section.

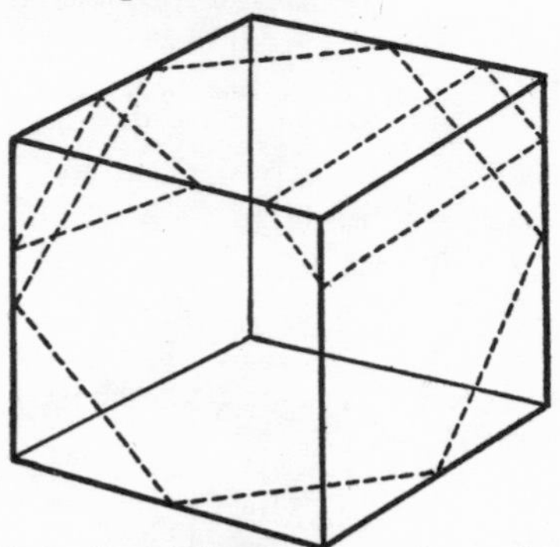

FIG. 8. *Diagram to show how sections of various shapes (three- four- and six-sided) can be cut from a cube.*

In a rock that is not too coarsely crystalline, however, there are generally sufficient grains of a particular mineral present, for some of them to show really characteristic and often diagnostic outlines. Numerous examples are quoted in the second part of this book.

Minerals differ greatly in the extent to which they show their ideal crystal form. Apatite might be quoted as one mineral that is generally well-formed, at least in igneous rocks, and the characteristic elongated prismatic or hexagonal outlines are often of great help in its identification. Quartz, on the other hand, is generally completely anhedral.

CLEAVAGE AND FRACTURE

Even a slight acquaintance with minerals as seen in hand specimen will be sufficient to enable the student to recall examples of minerals showing a tendency to break in definite planes. Calcite splits in three directions; in fact, it is almost impossible to break the mineral in any other direction. The common micas, muscovite and biotite, cleave in one direction with extreme facility; and gypsum also has one perfect cleavage. In thin section, these cleavages are seen as parallel straight lines. Minerals having more than one cleavage direction usually show more than one set of such parallel cracks, those of the one set intersecting the others. The student must realise that these cleavages have a definite crystallographic orientation; in the case of the micas, the one perfect cleavage is parallel to the basal pinacoid; the three cleavage directions of calcite are parallel to the faces of a rhombohedron.

It must be understood also that they are not necessarily parallel to faces actually present in the crystal. Calcite, for instance, crystallises in a bewildering variety of forms; but even the most complex of the latter break down into the simple shape controlled by the three sets of cleavage planes parallel to the unit rhombohedron ($10\bar{1}1$). In addition, of course, cleavage may be present in even the most irregular grains which exhibit no external crystal faces at all.

The appearance of a cleavage varies according to the orientation of the mineral grain in relation to the plane of the thin section. This can be illustrated by considering the case of mica. When the mica is lying flat, with its basal pinacoid in the plane of the section, no cleavage is visible;

although when it is lying with its pinacoid perpendicular to the plane of the section, mica exhibits a most perfect set of cleavage traces.

If we now pass on to the case of a mineral showing two cleavages, we have to consider, in addition to the presence of cleavage, the angle of intersection between the cleavages. Hornblende (monoclinic) is a common rock-forming mineral with two cleavage directions which are parallel to the prism faces (110).

A section cut at right angles to the cleavages shows the two sets of cleavage planes intersecting at angles of 124° and 56° (fig. 11). In a longitudinal or vertical section, however, the traces of the two cleavages appear as parallel lines. It is important to realise that the angle quoted for the intersection of the cleavages only applies to sections *exactly* at right angles to the cleavage planes. In a rock-slice it would only be by pure chance that a crystal of hornblende might have this particular orientation. In general the sections through the crystals are neither transverse nor longitudinal; but oblique to both. However, there are generally enough sections which approximate reasonably closely to the transverse direction for the angle of intersection of the cleavages to be used as a diagnostic character of hornblende. There are no other common rock-forming minerals with 'cleavage angles' close to 124°. Augite and other minerals of the pyroxene group have prismatic cleavage giving two sets of cleavage planes; but these intersect at angles of 93° and 87° in a transverse section, and therefore augite can easily be distinguished from hornblende solely on account of differences in cleavage.

Cleavages are only visible in a section when they are reasonably steeply inclined to the plane of the slice. When they are perpendicular to the latter ('vertical') they give very sharply defined cleavage traces; but when they are inclined at any angle of more than about 45° to the vertical then cleavage traces become too broad and diffuse to be any longer visible. For this reason, a front pinacoid section of a hornblende crystal appears to be devoid of cleavage;

the cleavage planes are lying too far from the vertical and too close to the horizontal plane of the section (fig. 11).

We can now appreciate that cleavages provide useful and often diagnostic criteria relating to minerals in thin-section. It is essential, however, to remember the effects of varying orientation on the appearance of any cleavage, and always to examine a number of differently orientated grains of each mineral before passing judgement on its cleavage characteristics.

In some minerals there is no regular cleavage, and any partings take the form of irregular fractures. Quartz is the commonest mineral that has absolutely no cleavage; but it has a conchoidal fracture similar to that of glass. Because of its lack of cleavage quartz can readily be distinguished from all the felspar minerals. The cleavage of olivine is very imperfect and inconspicuous; but this is another mineral which has a very prominent fracture, which is often accentuated by serpentinisation and the development of iron ore along the cracks.

REFRACTIVE INDEX AND SURFACE RELIEF

A brief explanation of refractive index has been given in Chapter 3. For a full account the student should refer to text-books on Light. For our present purposes we are only concerned with refractive indices of minerals in so far as they affect the appearance of the latter in thin-section.

A rock section is an extremely thin slab of rock which is sandwiched in between layers of a mounting medium, commonly a substance known as Canada balsam; this sandwich rests on a glass slide and above it is the thin cover glass. We are concerned, however, only with the mounting medium in contact with the minerals making up the rock.

It is well known that colourless transparent substances are visible only when they differ in optical density (refractive index) from the medium in which they are immersed; and the greater the difference between the two substances, the more conspicuous will be their limiting surfaces. An instructive experiment is to powder common glass and drop it into water. Better still, powder a fragment of the mineral cryolite and drop the powder into water; it is with the greatest difficulty that the mineral can be seen when immersed in the liquid, and one is inclined to believe that it has passed into solution. It is invisible, however, merely because the two substances, cryolite and water, have almost exactly the same refractive index.

We are now in a position to make a general statement. The visibility of a transparent solid depends on the difference in refractive index between it and the medium in which it is immersed. This medium may be a liquid, or in a thin-section of a rock it will be a colourless cementing material. Very commonly this cement is Canada balsam (R.I. = 1·54 approx.), which is a viscous liquid at ordinary temperatures, but sets hard after prolonged heating. On examining the margins of transparent minerals in section, we find, by working round the edges of the section where the minerals are in contact with Canada balsam, that some of the edges stand out conspicuously, while where other minerals come in contact with the balsam, the junction is comparatively inconspicuous. The minerals are in contact with the mounting medium also on their upper and lower surfaces, and these supply us with additional information. The surfaces are minutely pitted during the preparation of the section by the abrasive powders used. The roughness shows up when the refractive index of the mineral is far removed from that of the Canada balsam. The effect is known as **surface relief.** Minerals that have bold outlines because of their very high, or sometimes very low refractive indices, also show strong surface relief.

The Becke test. It is a simple matter to determine whether the refractive index is above or below that of the mounting medium by applying the Becke test. Using a high power objective, focus carefully an edge of the mineral in contact with the mounting medium, choosing an edge which is as clean as possible. If a small fragment of the mineral has become detached from the rest of the rock and thus lies surrounded by the medium, it will supply the ideal conditions. Rack the tube up and down a little on each side of the position of sharp focus. It will be seen that, when the fragment is slightly out of focus, a bright rim of light either surrounds or lies just within the mineral boundary. The bright line moves across the boundary as the tube is alternately raised or lowered. The only rule to remember is as follows: *when the tube is raised from a lower to a higher position, the bright line passes from the medium of lower to that of higher refractive index,* and, of course, conversely.

Certain precautions should be observed to obtain the best results from this test. The lighting and the mirror should be adjusted to give an absolutely central illumination. If the lighting is very bright it will be necessary to close the diaphragm somewhat to accentuate the Becke line. If the grain is too large or thick and its edges are irregular, stray reflections may give anomalous bright lines which behave contrary to the rule used in the Becke test.

The test may be used as a rapid means of distinguishing minerals of otherwise similar appearance whose refractive indices lie on opposite sides of that of the immersing medium, as for instance orthoclase (below) and quartz (above). It is also used in the actual measurements of refractive indices of minerals, which may be crushed and sieved, and small grains are immersed in a succession of liquids of different known refractive indices. Eventually a liquid may be found whose refractive index exactly corresponds with that of the mineral; or the refractive index of the mineral may lie between the closely-spaced indices of two liquids. In these circumstances the boundary of the grain almost disappears and the very faint and narrow bright line should ideally remain stationary when the focusing is altered.

The techniques of accurate refractive index measurement are essential to mineralogical research; but they are outside the scope of this book.

So far we have assumed that each mineral has a single refractive index, whereas we know that all anisotropic minerals are doubly refracting to greater or lesser extent.

This fact has to be taken into account in all accurate refractive index measurements; but in the routine examination of minerals it can be ignored except when the double refraction is very great. Calcite and the other carbonate minerals provide a case in point. The Becke line should be studied for *both* vibration directions when more complete information concerning the indices of a mineral grain are required.

'Twinkling.' Calcite is doubly refracting, and consequently, most of its sections possess the two rectangular vibration directions for transmitted light. When the stage is rotated, the mineral shows, in one position, a rough surface, well-defined borders, and conspicuous cleavages; in another position, a smooth surface, faintly defined borders, and inconspicuous cleavages. These extremes of relief are exhibited when the two vibration directions of the calcite in section come to be parallel in turn with the vibration direction of the light emerging from the polariser (E-W in fig. 9). This means that each of the two vibration directions has its own refractive index. A rapid rotation of the stage produces a rapid change of relief which cannot be better described than by the term 'twinkling'. Other minerals with a strong double refraction show the same effect, notably the other rhombohedral carbonates, but few other minerals show a conspicuous change of relief, though theoretically, all doubly-refracting sections ought to show some such

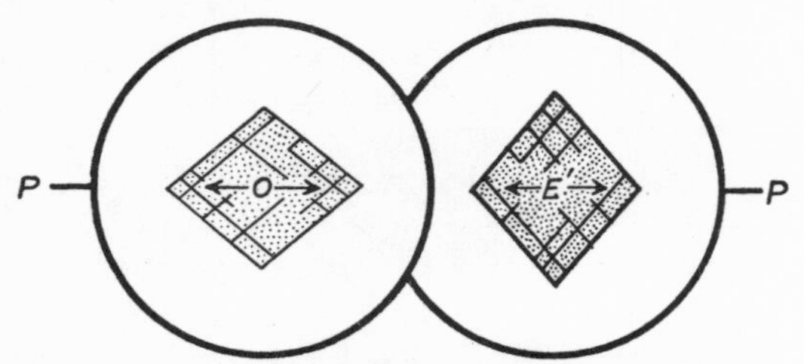

FIG. 9. *'Twinkling' effect in calcite. PP represents vibration direction of polarised light. O and E' represent the ordinary and extraordinary rays in cleavage fragments of calcite.*

change. Only in a few cases, however, such as the one cited, is the change sufficiently great to be worthy of consideration in practice.

COLOUR

Although the colour of minerals in hand specimen is often a most unreliable character for identification (as witness the milky, smoky, yellow, rose-coloured and amethystine varieties of quartz), the colours produced in thin section generally provide a much surer guide. A great number of minerals are quite colourless and transparent. Most of the common rock-forming minerals are either colourless or only very faintly coloured in section. Amongst the colourless ones are quartz, felspars, nepheline, leucite, olivine, cordierite, topaz, fluorite (generally) and many other less common minerals. Very faintly coloured ones include common garnet and augite. A few minerals are strongly absorbent and therefore dark coloured. Generally these are iron-bearing minerals. They include iron-spinel, andradite garnet and also minerals mentioned in the next section under the heading of pleochroism.

Pleochroism. Some anisotropic minerals show a distinct change of colour when viewed with plane polarised light and rotated on the microscope stage. This phenomenon is known as pleochroism. Biotite provides an excellent example of a pleochroic mineral. A section of biotite cut transverse to the cleavage (that is, a 'vertical' section) may show a change of colour from a pale straw yellow to a very dark brown. The extremes of colour are seen each time the cleavage lies parallel to one of the crosswires, or in other words a change takes place every 90° of rotation. If the vibration direction of the polariser is E-W, then the darkest colour occurs when the cleavage is in this direction, and the lightest colour when the cleavage is at right angles to the vibration plane of the polariser (fig. 10). The reason for the change in colour is quite simple to understand. Biotite is an anisotropic mineral and is therefore doubly refracting, so that light passing through it is split into two rays

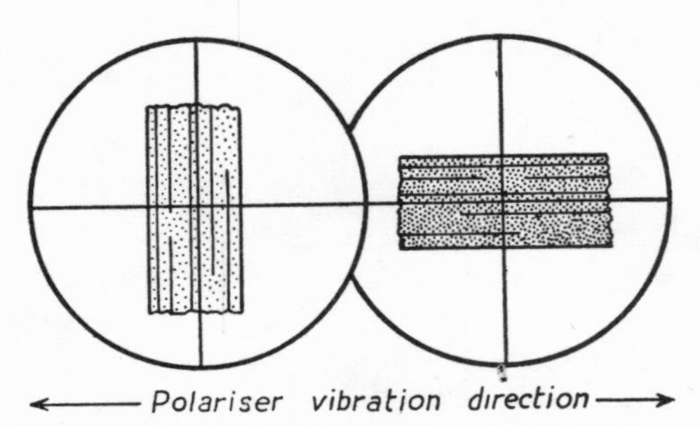

FIG. 10. *Pleochroism of mica. N.B. Polariser vibration direction may be N-S in some microscopes.*

vibrating at right angles to each other. The two rays suffer different amounts of absorption and therefore give different colours when they are viewed separately. With plane polarised light. this becomes possible, because each vibration direction of the mica can in turn be superimposed on that of the polariser, so that only the light vibrating in that direction is transmitted through the mica. When the mica section is rotated so that the cleavage is oblique to the crosswires, some part of both the rays and therefore of both the colours is seen simultaneously.

Without the aid of plane polarised light from a Nicol prism or polaroid disc, the eye cannot distinguish the two colours separately, because, with ordinary light both rays are transmitted by the mineral. This can be demonstrated very well with the aid of a small cube made from a crystal of coloured, say pink, tourmaline. It may be observed that, of the six faces of the cube, two parallel ones transmit a deep red colour, but the other four transmit a light pinkish tint. The two 'red' faces are parallel to the basal planes of the original crystal and are therefore perpendicular to the unique singly-refracting principal axis. The four light pink faces are parallel to this direction. If now the latter are viewed successively through a polaroid plate or nicol prism which is rotated against the cube face, the colour will be seen to change from the deep red already observed through the basal planes, to an almost colourless pinkish tint. The change of colour is, of course, due to the differential absorption of the plane-polarised light transmitted by the tourmaline and the polaroid in juxtaposition. The light pink seen by the naked eye is compounded of the two colours. Naturally with a tourmaline cube of a different colour, although the observed results would be different, they would be strictly comparable.

If in addition to the cube of tourmaline, one of glass or of a cubic mineral (for example pink fluorite) is also available, and is treated in the same way, no colour change occurs under any circumstances. This proves that pleochroism is a property confined to anisotropic minerals. Gemmologists can make use of this simple test to distinguish cut stones of ruby or sapphire (both anisotropic and pleochroic) from, say, glass of similar colour.

The phenomenon of pleochroism obviously supplies us with an extremely valuable means of distinguishing many minerals with the polarising microscope. Amongst common pleochroic minerals are the following:

Hornblende (pale green to dark green, or pale brown to dark brown);

Tourmaline (sometimes pale brown to dark brown; but with a very variable colour range);

Aegirine (rich yellow to emerald green);

Andalusite (colourless to an extremely pale pink).

Other examples are given in the second part of this book. It is important to realise that the colours depend very much upon composition and also upon the orientation of the mineral in the slice. Thus, not all biotites show the extreme range of colours from pale yellow to dark brown; some are greenish. Hornblendes and tourmalines are also very variable. One feature that does not change much, however, is the relative intensity of colour, or the amount of absorption shown by corresponding directions in differently coloured crystals of any mineral. Thus in any biotite, the vibration direction parallel with the cleavage always shows a much stronger absorption than the direction at right angles to the cleavage. In tourmaline, whatever the colours, it is always the vibrations parallel to the length of the prisms that are least absorbed and those across the prisms that show greatest absorption.

The complete determination of pleochroism and absorption of a mineral necessitates a knowledge of the principal vibration directions within the mineral. These are controlled by the atomic structure and symmetry of each type of crystal, as are all other optical properties. Thus Cubic minerals are non-pleochroic and show uniform absorption in all directions. Certain coloured Hexagonal, Trigonal and Tetragonal minerals are **dichroic,** that is, they show two colours corresponding to vibrations that are parallel to the *c*-axis (the extraordinary vibration) and at right angles to the *c*-axis (the ordinary vibration). Just as basal sections of Tetragonal, Trigonal and Hexagonal minerals are isotropic (p. 18), they also show no change of colour on rotation of the stage. Basal sections of tourmaline provide good examples for study. In Orthorhombic, Monoclinic and Triclinic coloured minerals conditions are more complex, and a complete statement of the pleochroism scheme is related to three principal vibration axes that are mutually at right angles to one another. A statement of pleochroism should therefore include three colours and three degrees of absorption for such minerals. For elementary purposes, however, it is sufficient to state whether a mineral is pleochroic, and to give the range of colours without referring them to particular optical directions.

Pleochroic Haloes. Certain minerals show curious circular coloured areas, each of which contains at its centre an included fragment of some other mineral, not necessarily visible in the plane of the slide. These coloured patches change colour when the section is rotated above the polariser; they are, in fact, pleochroic areas in an otherwise, possibly, colourless section. It is quite certain that the including mineral has acquired this property in consequence of the presence of such a minute inclusion. Several theories have been put forward to account for these pleochroic haloes, but it has now been satisfactorily proved that the development of colour around the inclusion is the result of the fact that the latter is radio-active, and that its ejected particles have produced an ionisation effect on the mineral in contact. Minerals showing these haloes include muscovite, cordierite, tourmaline, and biotite. The last two minerals are not colourless, even in thin section, but in their case the colour has been rendered more intense around the inclusion.

CRYSTALLOGRAPHIC SUMMARY

It is a matter of some difficulty to tabulate the pleochroic properties of minerals crystallographically in such a way as to be of assistance to the student, but it may be stated quite definitely that the following sections are non-pleochroic:

All sections of colourless minerals.
All sections of Cubic minerals.
Basal sections of Tetragonal, Hexagonal and Trigonal minerals.

Other sections may show pleochroism. Some show a striking change of colour, others merely a slight change; in still others the absorptive powers of the mineral for the two vibration directions are so nearly equal that even the most imaginative observer fails to notice any change of shade or colour.

CHAPTER 5

The Examination of Minerals between Crossed Polarisers

For many years the combined use of both polariser and analyser has been referred to as 'crossed nicols'. With the increasing use of polaroid the term is obviously inappropriate, and an alternative must be found. Since the function of the analyser is identical with that of the polariser—to produce plane-polarised light — it seems appropriate to adopt the term 'crossed polarisers'.* The word 'polariser' used alone will still refer to the lower unit, often formerly referred to as the 'lower nicol'.

When the two planes of polarisation are correctly crossed, light from the polariser is completely stopped by the analyser, so that darkness results when there is no mineral section interposed between them. The polariser, however, is in a rotatable mounting, and it may be that the two planes of polarisation are not at right angles. When this occurs, some light is passed by the analyser. *It is very important to check that the polarisers are always correctly crossed, as shown by complete darkening of the field.* Once the lower polariser has been turned to its correct position (sometimes indicated by a slight 'click') it should be left severely alone.

* The alternative term 'polar' is used in some recent books; but as this term is already used in crystallography in a totally different sense, it seems less appropriate than 'polariser'.

EFFECTS OF INSERTING A MINERAL SECTION BETWEEN CROSSED POLARISERS

When a mineral- or rock-section is interposed between polariser and analyser it is found that the majority of mineral sections affect the light in such a way that colours are produced. Further, rotation of the stage causes these colours to change in intensity, and in certain positions the sections may appear quite black, with complete extinction of all the light. The various effects seen may appear at first to be rather bewildering, and in order to sort them out, it is necessary to consider the phenomena one by one.

ISOTROPIC AND ANISOTROPIC MINERALS BETWEEN CROSSED POLARISERS

We may state at this point that the colours produced by minerals between crossed polarisers are due to the double refraction of the minerals concerned. All doubly refracting minerals are anisotropic, as explained on p. 16. Therefore all anisotropic sections of minerals are capable of producing colours seen between crossed polarisers.

On the other hand, isotropic (that is, singly-refracting) minerals *remain* black between crossed polarisers: turning the microscope stage has no effect upon them. Note that continuous blacking out of the light by one grain of a given mineral does not prove that mineral to be Cubic. Basal sections of Hexagonal, Trigonal and Tetragonal minerals are also isotropic (see fig. 4). On the other hand, if all the grains of a particular mineral in a rock slice appear black between crossed polarisers, that is certain proof of its Cubic nature. In practice the number of Cubic minerals commonly seen in rock-sections is very limited, and the student soon becomes familiar with them. They include common garnet, which is generally faintly coloured pink in plane polarised light; fluorite, characterised by a strong (negative) surface relief due to its very low refractive index; a form of leucite found in rapidly-cooled lavas and generally recognised by the rounded shape of its crystals; certain other rare felspathoidal minerals; and spinel, which is almost confined to certain metamorphic rocks.

Extinction. It is found that anisotropic sections transmit light of a certain colour in most positions of the stage, but from time to time during rotation they become dark. They are said to extinguish. If the angle of rotation is measured, it will be found that extinction takes place at intervals of 90° exactly.

We may consider apatite as a first example. Apatite, as seen in rock-sections, is a colourless Hexagonal mineral which normally occurs in prismatic crystals that show a rather high surface relief with plane polarised light. Six-sided basal sections are isotropic, and therefore always appear black between crossed polarisers. When a vertical prismatic section is rotated, however, the characteristic grey polarisation colour (see below) becomes extinguished when the length of the prism is lying parallel with one or other of the crosswires. In this example the extinction position is normally obtained by reference to the outer margins of the crystals. Muscovite is a good example of a mineral in which the excellent cleavage provides the reference direction against which the extinction is measured. Extinction occurs when the cleavage is parallel to either crosswire. In between these positions muscovite displays bright colours which are very distinctive. Both apatite and muscovite show **straight extinction** which occurs when a crystal margin or a cleavage is placed parallel to either the north-south or east-west crosswire.

In some sections of other minerals, however, extinction occurs when a cleavage or prominent crystal edge lies oblique to the crosswires. These sections provide examples of **oblique extinction.** An angle measured between the cleavage or edge of the crystal in extinction and the north-south crosswire is known as an **extinction angle.** It can be measured by noting the position of the stage when the mineral is in extinction, and again when the cleavage is parallel to the north-south crosswire, and subtracting one reading from the other (fig. 11).

Extinction occurs whenever one of the two rectangular planes of polarisation present in any anisotropic mineral section, lies parallel to the vibration plane of the polariser (see fig. 4). When this happens the light from the polariser passes through the mineral slice without any deflection, and is, of course, completely eliminated at the analyser. It is easy to see now why extinction occurs four times in a complete rotation of the stage.

Provided measurements of extinction angles are made with discretion and from well-chosen sections, they can provide a most valuable aid to mineral identification. The choice of a suitable grain from the many that may be present in a rock must be guided by a knowledge of the symmetry of the mineral concerned. Reference to fig. 4 will help the student at this point. Straight extinction occurs in every case in which the two vibration directions are parallel to appropriate crystal edges or cleavage traces. It is applicable to strictly vertical sections of Tetragonal, Hexagonal and Trigonal minerals. One of the two vibrations is parallel to the vertical axis of the mineral and the other is perpendicular to it. The former is the extraordinary, and the latter the ordinary ray. Sections of Orthorhombic minerals cut parallel to any one of the crystal axes—this includes all pinacoidal and prismatic sections—also show straight extinction. In Monoclinic minerals such as augite and hornblende the only sections which can show straight extinction are those perpendicular to the side (clino-) pinacoid. In Triclinic crystals all sections normally extinguish obliquely.

Two points deserve special mention since they provide stumbling blocks for beginners. The first concerns the so-called straight extinction of Orthorhombic minerals. It is only when sections of such minerals are cut *strictly parallel* to a crystal axis that the vibration directions are correctly orientated so as to result in straight extinction. Any degree of random obliquity of the slice will cause the extinction to be oblique in some degree. Often such oblique sections can be detected because cleavage traces tend to lose their sharp definition, and in the case of euhedral crystals, the shapes will appear distorted and assymmetrical.

Strictly, this argument applies also to oblique sections of Tetragonal, Hexagonal and Trigonal minerals: a section somewhat oblique to the vertical axis of a Hexagonal prism will obviously have non-parallel sides.

The second stumbling block concerns the use of extinction angles for identifying minerals, particularly those crystallising in the Monoclinic System. As explained above, most of such sections show oblique extinction to some degree; but it is only the *maximum angle* which is of the slightest diagnostic value. Hornblende is a case in point

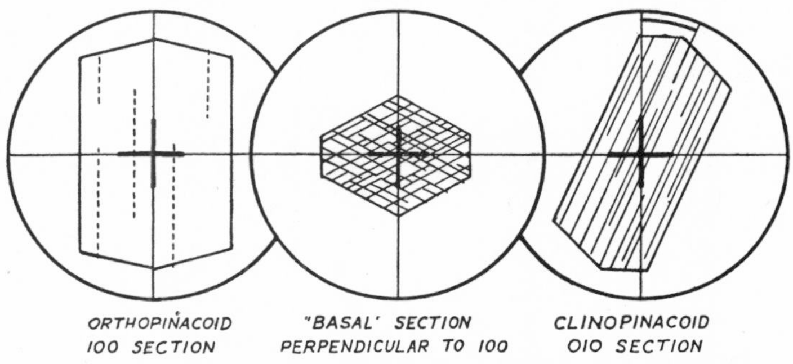

FIG. 11. *Diagram to show the straight, symmetrical and oblique extinction of sections of a monoclinic amphibole. Vibration directions are indicated by the thick lines parallel to crosswires. Cleavage in the* (100) *section is shown by broken lines since it is too inclined to the plane of section to be visible.*

(fig. 11). The front pinacoid section shows straight extinction, as demanded by the crystallographic and optical symmetry. A section cut parallel to the side (clino-) pinacoid shows oblique extinction at the maximum angle of 25° for this mineral. Any section in zone with these two, but occupying an intermediate position, will show oblique extinction at any angle between 0° and 25°. It is possible to find sections showing an extinction angle greater than the theoretical maximum by introducing another direction of tilt. Thus a pyramidal section, inclined to all three crystallographic axes, could have an extinction position widely oblique to the visible cleavage traces.

Interference or Polarisation Colours. When an anisotropic mineral section is lying with its two vibration planes inclined to the vibration directions of the crossed polarisers, the section appears coloured. The colours are produced in between positions of extinction therefore, and reach their maximum intensity midway between two extinction positions. The writer generally refers to such positions as the *45-degree positions* for the vibration planes of a mineral plate. The colours vary greatly according to the nature and orientation of each mineral, and they may range from dull grey through shades of yellow, bright orange, red, blue and green to watery pinks and very pale greens. These are the colours of **Newton's Scale** shown in the frontispiece of this book. Grey is referred to as a low order colour, and the pale pinks and greens, which grade into white, belong to high orders. It will be seen that there is an almost exact repetition of colour sequences (though not of intensity) in the first three orders. In general the most intense colours of the whole scale are those of the second order. The interference or polarisation colours of most minerals seen in normal rock-sections fall within these first three orders.

Before entering into a detailed explanation of the phenomenon, it will be helpful to make an elementary statement concerning the nature of interference colours.

1. Interference colours are caused by the double refraction of anisotropic minerals. The greater the maximum double refraction (*i.e.* difference between the extreme refractive indices) of a mineral, the higher is the order of the colour produced.

2. For a given mineral the colours depend upon the orientation of different grains. Thus in a mineral like apatite (Hexagonal) a section cut parallel to the vertical crystallographic axis shows the highest order colour which is actually a light grey, while basal sections are isotropic.

Sections intermediate between the vertical and basal show various darker shades of grey. In a mineral description, only the highest order colours yielded by the mineral are referred to. Thus the normal interference colour of apatite would be given as 'light grey', not as 'grey to black'.

3. *The order of colour depends upon the thickness of the section or grain.* A standard thickness of about 0·03 mm. has been adopted so that all minerals may be viewed under similar conditions. In the preparation of a rock-section, the grinding process is checked at intervals by noting the interference colours of certain standard minerals such as quartz or felspar. When these minerals show colours which range up to greyish-white or pale yellow, grinding is stopped. Some old or badly-made slides can be distinguished by the garish colours produced by a mineral such as quartz which actually has a weak double refraction and which should therefore yield colours low down in the first order.

For the general student, items 1 and 2 above are of greatest importance. The thickness of a section (item 3) can usually be regarded as a fixed quantity. The following table shows the relationship between the value of the double refraction and the interference colours produced by a number of minerals. In all the minerals listed except quartz, isomorphous substitution causes considerable variation in optical properties, including the birefringence, so that the values given must only be regarded as representative. Hornblende and biotite show very wide variation.

TABLE

Mineral	*Double refraction*	*Maximum colour and order (in 0·03 mm. section)*	
Leucite	0·001	Dark grey	First
Nepheline	0·004	Grey	,,
Apatite	0·005	Grey	,,
Orthoclase	0·007	Light grey	,,
Quartz	0·009	Yellowish white	,,

Mineral		*Double refraction*		*Maximum colour and order (in 0·03 mm. section)*	
Hornblende	..	0·020	..	Purple	Second
Augite	..	0·025	..	Green	,,
Olivine	..	0·038	..	Blue	Third
Muscovite	..	0·038	..	Blue	,,
Biotite	..	0·045	..	Yellowish green	,,
Sphene	..	0·140	..	Near white	Fifth
Calcite	..	0·169	..	Near white	Sixth

Item 2, that is, the effect of varying the orientation of crystals, can be demonstrated with any rock-section composed of only one mineral, such as quartzite, or dunite (an igneous rock composed almost entirely of olivine). Since the sections are of uniform thickness, the various colours seen can only be due to the orientation of the grains. Under these conditions olivine crystals in a dunite section may show *every* colour in Newton's Scale from third order blue to first order grey. Some grains will be in positions of extinction. The effects of extinction are transitory, however, and should never be confused with the permanent effect of the three-dimensional orientation of each grain in the rock-slice. The student should prove for himself that normally the interference colour of a particular crystal cannot be altered by rotation of the stage: only the intensity of the colour varies.

It must be clearly understood that, for a constant thickness, every birefringent mineral will exhibit a definite *range* of interference colours—a wide or a narrow range in proportion to the specific birefringence of the mineral. There is no particular interference colour that is characteristic of a particular mineral: on the other hand *some* sections of several different minerals must exhibit precisely the same interference colour.

The best demonstration of the effect of variable thickness (item 3) is provided by a **quartz-wedge** (see p. 56). This is a gently tapered, slender wedge cut from a single quartz crystal and polished. If this is gradually inserted into the field of view, and always held so that the length of the wedge is at about 45° to the crosswires, a succession of colour bands will appear in the sequence shown on Newton's Scale. Ideally the colours would start with black where the thickness of the wedge is zero, and end up with anything from 4th to 7th order colours according to the maximum thickness of the wedge.

One further point deserves mention. When a mineral is strongly coloured in plane polarised light, the interference colours may be almost unrecognisable through the masking effect of the body colour. This is well illustrated by biotite whose dark brown absorption colour may effectively obscure the high-order interference colours. Muscovite, on the other hand, lacking any absorption, shows similar bright interference colours to perfection. Sphene provides another interesting case in which the interference colours are of such a high order, and therefore so pale, that they are scarcely visible through the golden-brown or grey-brown body-colour. This is one of the very few minerals, therefore, which does not change in appearance when the polarisers are crossed, provided the section is lying in the 45° position (*i.e.* it is not in extinction).

There are several special features displayed by minerals between crossed polarisers which are caused by peculiarities in their mode of crystallisation. Among these phenomena are zoning and twinning. Although in one sense they provide added complications for the student to master, in another they yield very useful evidence relating to the histories of the minerals concerned, and zoning and twinning provide extra criteria for distinguishing minerals as described below.

ZONING

Any crystal found in an igneous rock will have formed gradually over a considerable range of falling temperature. Igneous rock-forming minerals belong, in the main, to isomorphous series in each of which there is a possibility of a certain amount of atomic substitution. Crystallisation commences in every case with precipitation of a high-temperature member of the series, and as cooling proceeds there is a constant interchange of atoms between the growing crystal and the surrounding liquid. If cooling is very slow, the whole crystal remains homogenous and maintains equilibrium with the liquid. At the close of crystallisation the composition achieved is that of a lower-temperature member of the series. When cooling is more rapid, atomic substitution does not keep pace with the rate of growth of the crystal which therefore becomes zoned. As a result successive layers of crystal growth have varying compositions. Since optical properties are controlled by composition it follows that these show variation from the cores of zoned crystals to their margins. The zoning may affect the extinction positions as in the case of plagioclase. Between crossed polarisers zoned plagioclase crystals display concentric shells of varying extinction and therefore showing varying shades of grey interference colours. This feature is especially conspicuous in plagioclase phenocrysts in rapidly cooled lavas.

Zoning may be apparent in other ways, as for instance variation in the colour displayed by a crystal seen in plane polarised light. This is a feature of some varieties of pyroxene for example, in which the outermost zones are enriched in sodium, giving greenish rims (of aegirine-augite) around more or less colourless or light brown cores. Tourmaline, garnet and sphene are other minerals which often show colour-zoning. The phenomenon is not confined to 'igneous' minerals, however, and zoning may be conspicuous, for example, in some varieties of metamorphic garnet.

Occasionally successive stages of crystal-growth may be marked by concentrations of minute inclusions of some foreign minerals; or they may be picked out by selective alteration of certain zones. The student should always be on the look-out for details of this kind, for in them may be found some of the items of evidence by means of which the complex history of a rock may be unravelled.

TWINNING

Observant students will have noticed that grains or crystals of certain minerals which appear as single crystallographic units in plane polarised light show sharply divided areas of contrasted interference colours between crossed polarisers. The separate areas are generally divided by straight boundaries. On rotating the stage it is found that they also have different extinction positions. These are the characteristics of twinned crystals. Twinning produces rotation or reversal of the crystal structure, the change-over taking place at a well-defined plane (the twin plane) in that structure. It follows, therefore, that the optical characters should be similarly reversed at each twin-plane.

Examination between crossed polarisers often provides the surest, and sometimes the only, guide to the presence of twinning, even when the external form of a crystal gives no indication of it. A good example of this is provided by the relatively low-temperature variety of leucite (see p. 115). Crystals of leucite are often well-formed icositetrahedra, one of the more complex forms belonging to the Cubic system. In section and between crossed polarisers, however, such crystals are slightly anisotropic and show frequently-repeated twinning in several directions (see Pl. IX). Strong illumination is needed to see the twinning to advantage, and then it is very striking. Twinning provides a diagnostic feature of a number of minerals and is particularly useful in distinguishing the various kinds of felspar (p. 110). In orthoclase, the commonest type of twin (twinned on the Carlsbad Law) produces crystals that are divided down their length (parallel to 010).* The two halves show different tones of grey interference colour and extinguish in different positions in most sections. When the twin-plane divides the crystals into two parts as in orthoclase, the twinning is said to be simple.

* This is known as the composition plane of the twinned crystal. The twin plane in this case is actually parallel to (100).

Plagioclase shows multiple, lamellar twinning (on the Albite and Pericline Laws). In section the alternating lamellæ appear in contrasting shades of grey, or of black when one set of lamellæ is in extinction. An analogy might be made with the structure of ply-wood in which the grain of the wood (comparable with the atomic structure of plagioclase) is reversed in orientation in alternate lamellæ. Generally the presence of multiple twinning removes the identification of plagioclase beyond doubt. However, a particular grain may happen to be sectioned parallel to the twin-planes, and so fails to show the characteristic lamellæ. Its differentiation from orthoclase must then rest upon comparison with other lamellar-twinned crystals in the slice, upon its degree of alteration, and its refractive indices.

Microcline is a Triclinic form of potassic felspar, having the same composition as orthoclase. Being Triclinic it has twinning on the same general laws as the plagioclase felspars. The lamellæ are usually more slender than those of plagioclase, and are generally spindle-shaped. Further, two sets of lamellæ usually intersect one another at approximately 90°, giving a cross-hatched effect. Thus we see that the three main kinds of felspar can be distinguished at a glance by their twinning.

Other minerals which frequently display twinning are pyroxenes and amphiboles. In both cases it is instructive to note how the prismatic cleavages extend across the (100) twin planes without interruption.

CHAPTER 6

Use of the Microscope Accessories and Observations in Convergent Light

---

FORMATION OF INTERFERENCE COLOURS

Fundamentally interference colours are produced because one ray emerging from a mineral slice is retarded behind its faster counterpart. The amount of the retardation, R, can be calculated with great ease. A very simple analogy is provided by individuals in a race running at different speeds over a certain track. The longer the track the greater will be the distance (or retardation) separating the runners at the end of the race. The track length corresponds to the thickness, t, of a mineral slice or of a grain and the two speeds correspond to the two refractive indices, n_1 and n_2, of the section. We then have:

$$R = t(n_1 - n_2)$$

It is necessary, of course, to use measurements of length which are commensurate with the extremely minute wave lengths of light, and for this purpose the units chosen are millionths of a millimetre, $\mu\mu$. Retardation expressed in these units is sometimes referred to as the **colour value,** because each retardation corresponds to a particular colour as shown in fig. 12. This diagram is extremely useful, especially if it is used with the colour scale forming the Frontispiece. Fig. 12 is, of course, merely a graphical representation of the equation given above. Its use can be illustrated in the solution of two simple problems.

1. To estimate the thickness of a section containing quartz which shows pale yellow as its highest order colour.

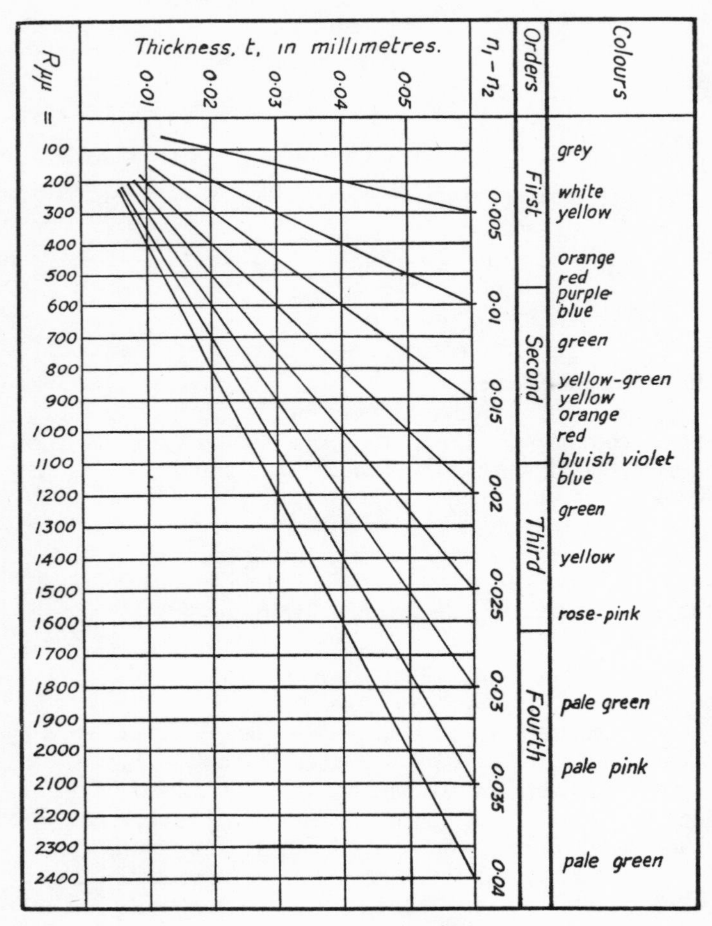

FIG. 12. *Birefringence chart.*

From the chart, R is estimated to have a colour value of about 270. The maximum double refraction ($n_1 - n_2$) of quartz is known to be 0·009. Because quartz has a very uniform composition, this value is fixed and provides one of the most useful standards in crystal optics. Deriving the thickness of section by calculation we have:

$$270 = t \times 0{\cdot}009$$
$$\therefore \quad t = 30{,}000 \ \mu\mu$$

or more conveniently, 0·03 millimetres.* The same result can be obtained from the chart by running the eye down the vertical line corresponding to a colour value of 270 until it intersects the sloping line corresponding with a birefringence of 0·009. The intersection occurs on the horizontal line indicating the thickness of 0·03 mm.

2. In the same section an unknown mineral yields a maximum interference colour of third order blue corresponding to a colour value of approximately 1150. What is its birefringence? From the information obtained above, we know the thickness is 0·03 mm., or 30 microns, or 30,000 $\mu\mu$.

$$1150 = 30{,}000\ (n_1 - n_2)$$
$$\therefore n_1 - n_2 = 0{\cdot}036.$$

From the chart a value of 0·035+ would have been obtained by tracing the appropriate line of birefringence, from the point where the lines of thickness and colour-value intersect, diagonally up to the top of the diagram.

DETERMINATION OF OPTICAL SIGN WITH THE AID OF ACCESSORY PLATES

It is sometimes necessary to know which of the two rays (the ordinary and the extraordinary) in a mineral section is the faster. This information is needed, for instance, whenever a complete optical examination of a mineral is being undertaken as described below. The technique is simple. The mineral fragment under examination is rotated to its 45-degree position. Its interference colour is noted. Then an accessory plate—which is really another mineral slice of

* The unit of measurement used may be a thousandth part of a millimetre, or micron, μ. The thickness quoted is therefore 30 microns, which is standard for a good quality rock-section.

known characteristics—is superimposed, also in the 45-degree position. The combined effects of the mineral and the accessory yield a second interference colour which is either of higher or lower order than the first.

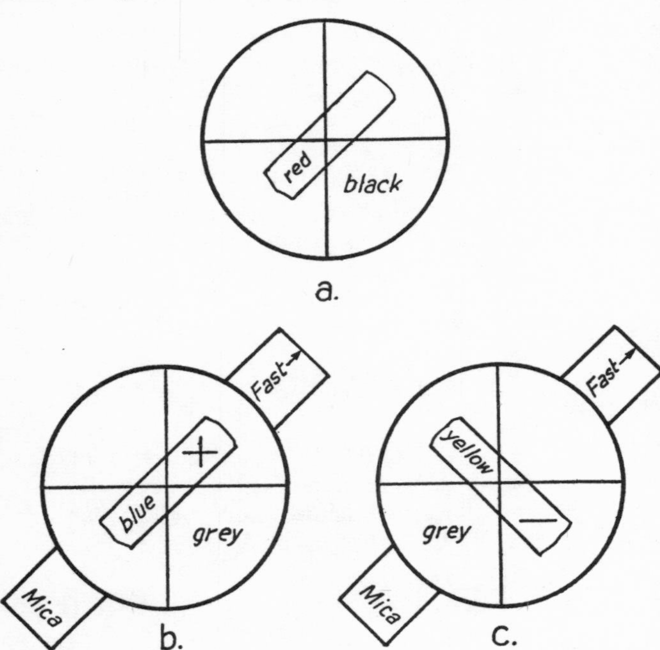

FIG. 13. *Use of a mica plate to determine sign of elongation of a prismatic mineral. The + and — signs indicate addition and subtraction of interference colour value.*

(1) Interference colour increased. This occurs when *like* vibrations are superimposed, with the fast and slow rays of the accessory plate lying in the same planes as the fast and slow rays of the mineral.

(2) Interference colours decreased. This occurs when *unlike* vibrations are superimposed, with the fast ray of the accessory on the slow ray of the mineral and vice versa.

Three kinds of accessories are in common use: *viz.* the mica plate, gypsum plate and the quartz-wedge already mentioned (p. 46). They are in specially constructed glass, metal or plastic mountings which can be inserted into a slot cut into the microscope tube above the objective. The slot ensures that the accessories lie with an elongation at 45 degrees to the crosswires. The fast and slow directions should be engraved on the accessories. Some makes are length-fast and others length-slow.

A *mica plate* is prepared from a flake of white mica cleaved to such a thickness as to produce a pale grey interference colour corresponding to a colour value of about 150. This small retardation is especially valuable for testing minerals which give bright interference colours of second, third and fourth orders. Suppose that a suitably orientated section of a mineral yields a second order red (colour value 1050). The colour produced with a superimposed mica plate will be either blue of a higher order (colour value: 1050 + 150 = 1200), or yellow (colour value 1050 − 150 = 900), according to whether like or unlike vibrations are superimposed. The colour changes are shown in Fig. 13, a, b and c respectively. In this example the unknown mineral is proved to be length-fast.

A *gypsum plate* is designed to give a purple interference colour, referred to as the **sensitive purple** (colour value approximately 570), which divides the first and second order colours. The term 'sensitive' is used because the addition or subtraction of a small extra retardation alters the colour in a very striking way. A gypsum plate is therefore used in conjunction with minerals of weak double refraction, and especially with those which normally produce grey interference colours. Suppose a mineral section is grey corresponding to a retardation of 200. Superimposing a gypsum plate will make the colour change to green (570 + 200) or to an orange-yellow (570 − 200).

Operation of accessory plates can be explained by extending the analogy with the runners mentioned on p. 51. This time they must be pictured as taking part in a relay race. If two slow runners are in one team and two fast ones in the other the net retardation will be increased by the end of the race. Conversely, if a fast runner succeeds the slow one of the first lap and vice versa, the retardation at the end of the race will be reduced.

A *quartz-wedge* operates in exactly the same way; but it has the additional advantage that it covers several complete orders of colour. A wedge is ideal for use with isolated mineral grains set in a mounting medium such as Canada balsam. As the wedge is slowly inserted into the 45-degree slot, it colours the whole field of view with the sequence of colour bands of Newton's Scale. The mineral grain (also lying in its 45-degree position) will contribute its own retardation. The colour value of the grain will either be increased or decreased relative to the colour value of the quartz-wedge seen in the surrounding field. If it is decreased there will come a point where the retardation of the grain is exactly balanced by that of the quartz-wedge. **Compensation** is then achieved, and ideally the colour of the grain should be completely extinguished.

The production of compensation can be used in two ways. Firstly, it can only occur when unlike vibrations of mineral and quartz-wedge are parallel, and therefore provides a guide to the **optical sign** of the mineral. Secondly, it can be used to obtain the true colour value of middle and high order interference colours. Second and third order blues, greens and yellows are very similar and may be confused. This difficulty is overcome if a record is kept of the colours produced by the wedge up to the point of compensation.

THE RELATIONSHIP BETWEEN CRYSTALLOGRAPHIC AND OPTICAL SYMMETRY

A glance at fig. 4 provides a reminder that minerals can be divided into seven crystal systems (though the Trigonal representative is not shown); but there are only three main subdivisions that can be made on the basis of optical symmetry. These are:

1. *Isotropic.* Vibrations take place with equal ease in all directions. Only Cubic minerals are completely isotropic.
2. *Uniaxial.* Light transmitted along the unique c-crystallographic axis vibrates with equal ease in all directions at right angles to this axis. Basal sections of Tetragonal, Trigonal and Hexagonal minerals are therefore isotropic. All other sections are anisotropic. The isotropic c-axis direction is called the **optic axis.**
3. *Biaxial.* Minerals of the remaining crystal systems, the Orthorhombic, Monoclinic and Triclinic, lack this simple optical symmetry. It would appear from the diagram (fig. 4) that all sections of these remaining minerals are anisotropic. Actually each one of these minerals has two isotropic directions known as the two optic axes. Determining the positions of optic axes requires apparatus and techniques which are well beyond the scope of the beginner. All that needs to be known at this stage is the fact that two optic axes exist in biaxial minerals, and that these minerals can be rapidly distinguished from the uniaxial ones by tests described below.

SIGN AND OPTICAL ORIENTATION OF UNIAXIAL MINERALS

Basal sections of uniaxial minerals are isotropic. In contrast to this, vertical sections display the maximum order of interference colour appropriate to each mineral because such sections contain the fastest and slowest vibration directions. The description of apatite given on p. 44 emphasises this point.

The *slowest* vibration direction is called the Z direction. This gives rise to the maximum refractive index of the mineral, referred to as $\mathbf{n}\gamma$ or $\mathbf{n}_z$, or merely γ.

The *fastest* vibration direction is called the X direction. This gives rise to the minimum refractive index, $\mathbf{n}\alpha$ or $\mathbf{n}_x$

When Z is parallel to the c-axis of the mineral, the latter is said to be **optically positive.** When X occupies this position the mineral is **optically negative.**

In order to determine the sign, it is only necessary to locate a *vertical* section and to use a suitable accessory plate in the manner described above (p. 55). It is easy enough to find a suitable vertical section when the mineral has an elongated prismatic habit as is the case with apatite and tourmaline. Quartz and calcite, however, generally occur in granular aggregates which give no indication of the position of the vertical *c*-axis in each grain. If the sign of such a mineral is required it must be obtained by using convergent light, as described on p. 62.

OPTICAL ORIENTATION OF BIAXIAL MINERALS

We have seen that there are two vibration directions in all anisotropic sections of minerals. In the case of uniaxial minerals it was found necessary to distinguish two principal directions of vibration, namely the vibration parallel to the *c*-axis, and that perpendicular to the *c*-axis of each mineral. In biaxial minerals it is necessary to recognise *three* principal vibration directions out of the infinite number of possible vibrations present in the mineral. Two correspond to the fastest and slowest directions, X and Z. The third has an intermediate velocity, Y, and a refractive index $\mathbf{n}\beta$ or $\mathbf{n}_{Y}$. *All three directions are mutually at right angles.* The position of these three principal vibrations in a mineral is controlled by crystallographic symmetry. In each of the three crystal systems — Orthorhombic, Monoclinic and Triclinic — the arrangement is different. In Orthorhombic minerals the three vibration directions are parallel to the crystallographic axes. In Monoclinic minerals *one* of the vibrations is always parallel to the b-crystallographic axis, and the remaining two lie within the (010) plane of symmetry. No general statement can be made concerning Triclinic minerals because of their lack of planes and axes of symmetry. Every Triclinic substance therefore has its own unique arrangement for the position of the vibration directions.

The information given above should be compared with the data given concerning extinction on p. 41 and in fig. 4.

Determining the complete optical orientation of a biaxial mineral—including the vibrations X, Y and Z, the position of the optic axes, and the optical sign—requires research techniques that are generally beyond the scope of the beginner. As an indication of how the principal vibration directions can be determined, we will describe a procedure relating to the Orthorhombic mineral, barytes (or barite).

Ideally, three pinacoidal sections should be available from a well-formed crystal. The excellent cleavages of barytes make these difficult to obtain; but on the other hand the cleavage traces provide indications of the directions of the *a, b* and *c* crystallographic axes in thin-section. Because barytes is Orthorhombic we know that the three principal vibrations must lie parallel to the axes *a, b* and *c*. There are six possible cases:

1	2	3	4	5	6
$a = X$	$a = X$	$a = Y$	$a = Y$	$a = Z$	$a = Z$
$b = Y$	$b = Z$	$b = X$	$b = Z$	$b = X$	$b = Y$
$c = Z$	$c = Y$	$c = Z$	$c = X$	$c = Y$	$c = X$

The correct interpretation for barytes is obtained by examining the three sections between crossed polarisers and using an appropriate accessory plate with each in turn. In this way the fast and slow vibrations in each pinacoid can be determined. The vibrations, of course, are parallel to the crosswires when each section is in a position of extinction. The section is then turned through 45 degrees and the accessory plate is superimposed.

It is found that:

In the (100) section, the vibration parallel to *c* is faster than that parallel to *b*.

In the (010) section, the vibration parallel to *c* is faster than the one parallel to *a*.

In the (001) section, the vibration parallel to *b* is faster than that parallel to *a*.

Hence it follows that *c* is faster than *b* which is faster than *a*.

Therefore: $a = Z$
$b = Y$
$c = X.$

Further, as the optic axial plane is, by definition, that plane which contains the fastest (X) and the slowest (Z) vibrations, we have shown that it must lie parallel to (010) as shown in fig. 14.

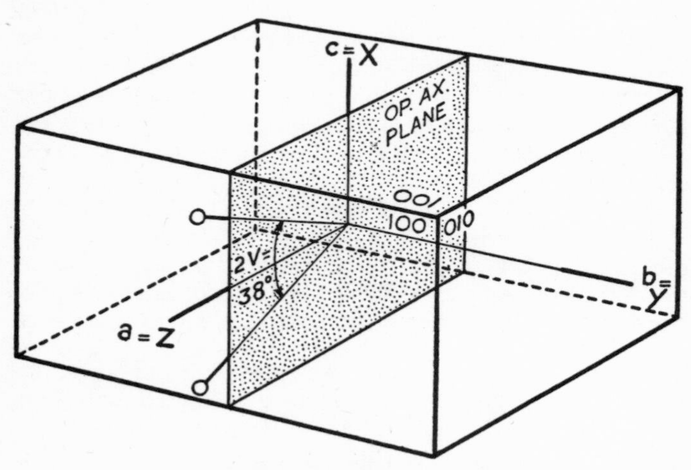

FIG. 14. *Optical orientation of barytes.*

If the three sections are of equal thickness, further valuable information can be obtained concerning their birefringences. The (010) section will show the highest order colour—and incidentally this is the highest possible colour value that is shown by any section of the mineral. The high colour follows from the fact that the fastest and slowest vibrations, X and Z, corresponding to the minimum and maximum refractive indices α and γ respectively, both lie in this plane. The basal section (001) gives colours that are of almost equally high order, while those of the front pinacoid (100) are much lower. The complete orientation of barytes is shown in Fig. 14.

A similar procedure can be adopted for Monoclinic minerals, though an added complication arises from the fact that only one of the principal vibrations lies parallel to a crystallographic axis. In a great many cases it happens that the Y-vibration is the one that is parallel to *b*. X and Z generally lie in the (010) plane and their positions can be obtained from the extinction positions of side-pinacoid sections. Confirmation of this relationship is obtained when the (010) sections show the highest order colours appropriate to the mineral. Minerals with this type of orientation include gypsum, sanidine, most pyroxenes and amphiboles.

The reader will have realised that assembling all the facts necessary for the complete optical orientation of a mineral involves expenditure of considerable time, and also needs rather favourable conditions of cleavage or external form by means of which the various critical sections can be recognised. For a uniaxial mineral it is only necessary to determine the character of the vibration parallel to the *c*-axis in order to obtain the optical sign. Even so, there are some uniaxial minerals which give no indication of the position of the *c*-axis. Granular quartz has already been mentioned in this connection. With biaxial minerals the problem is much more complex, and even when the positions of the X, Y and Z vibrations have been located, this still gives no indication of the optic sign. This last problem and many other problems of orientation, optical character (that is uniaxial or biaxial) and optical sign can be solved rapidly and easily by examining **interference figures** produced when the minerals are examined between crossed polarisers with convergent light instead of the parallel beam normally used.

METHODS OF OBTAINING AN INTERFERENCE FIGURE

In fitting up the microscope for convergent light, the instrument is first arranged as for use with parallel polarised light. A high-power objective (say, ×40, or a ¼ in.) must be used. The convergent light is produced by inserting a small hemispherical lens immediately below the stage so that the upper surface of this lens—the **converger**—is almost touching the glass object slide. Mechanisms for placing the converger in position vary according to the make of instrument. It may be permanently in position; or fixed in a metal slider attached to the stage; or on a swing-arm operated by a lever beneath the stage. It is important to have well-centred illumination, and absolutely essential to have the high-power objective centred so that the grain under examination does not depart at all from the centre of the field.

Under these optimum conditions it would appear that the only effect of the converger is to concentrate the light on the centre of the field. Actually, however, a small interference figure or optic picture is produced in a focal plane just above the objective, and this picture can be made visible by one of two processes. Either the eye-piece may be removed so that the observer can look straight down the tube at this picture, which then appears very small; or an additional lens, the **Bertrand lens,** can be inserted into the microscope tube beneath the eye-piece. This removes the ordinary magnified image from the field of view and brings the interference figure into focus. A Bertrand lens is supplied with more expensive instruments. It gives a larger interference figure; but the figure is not so bright and sharp as it is when viewed directly with the eye-piece removed. A student should never regard the absence of a Bertrand lens as any deterrent in obtaining perfectly adequate and workable figures.

The clarity of a figure can often be improved by making slight adjustments to the diaphragm beneath the stage, and also to the focusing. In some of the more expensive microscopes, the Bertrand lens is also adjustable, and may be fitted with an iris diaphragm.

COMPARISON OF UNIAXIAL AND BIAXIAL INTERFERENCE FIGURES

Every section of a doubly refracting mineral produces some kind of interference figure with convergent light. The amount of information that can be gleaned from a figure depends, however, upon the orientation of the section. The most informative figures are those which show the points of emergence of optic axes, and a selection of such figures is shown in Plate I. It will be seen that an interference figure consists of colour bands which are generally elliptical or circular in shape; and of black or shadowy **isogyres.** The latter may form a cross, or they may be curved in the shape of hyperbolæ. A fundamental distinction is made between uniaxial and biaxial interference figures based on the behaviour of the isogyres as described immediately below. Before proceeding further, however, the reader will be well advised to examine interference figures provided for instance by quartz, calcite and cleavage flakes of mica. These minerals give a wide diversity of figures and a critical examination of them will be found more informative than any number of diagrams.

(1) In a *centred uniaxial figure,* the isogyres form a cross which remains stationary as the stage is turned. The colour bands form concentric circles. The centre of the black cross corresponds to the point where the optic axis emerges. In other words the single optic axis (which is also the *c*-axis of the mineral) is perpendicular to the plane of the section and is parallel to the tube of the microscope. We already know that this is the isotropic direction for uniaxial minerals. *Thus, if a centred interference figure of a uniaxial mineral is required, search must be made for a grain which approaches as close as possible to an isotropic condition.*

Thus in a quartz rock, grains which polarise in the darkest greys or, ideally, remain in permanent extinction, provide the best interference figures.

(2) In the most useful type of *centred biaxial figure,* both optic axes emerge within the field of view. Where they emerge, they form pivot-points, as it were, about which two separate isogyres rotate. As the stage is rotated the two isogyres fuse together to give a cross which is not unlike that of a uniaxial figure; but further rotation causes them to separate and to become either slightly or very considerably curved according to the nature of the mineral. The colour bands vary somewhat in shape according to the mineral: they may form concentric circles around the individual pivot-points (known as the melatopes) of each isogyre, or form ellipses which enclose both of these points. An ideal substance to examine for a perfectly centred biaxial figure is a cleavage flake of muscovite.

INTERPRETATION OF UNIAXIAL FIGURES

We will start with the assumption that the figure is centred. A cone of light can be visualised as passing through the mineral plate and converging on the point where the isogyres cross. Some of the rays forming the cone will pass through the mineral in the E-W and N-S planes corresponding to the vibration planes of the polariser and analyser. These rays are extinguished and give rise to the crossing isogyres. Between the isogyres the rays approach the ideal 45-degree positions which yield interference colours. Every one of these is polarised in the usual manner into two planes. The planes are radial and tangential in relation to the cone of rays, as shown in fig. 15a. The radial vibrations, in planes which pass through the optic axis, are those of extraordinary rays (e′). Tangential vibration planes correspond to ordinary rays (o). The sign of a uniaxial mineral depends upon the relative velocities of these two vibrations. *When the ordinary ray is faster than the extraordinary, the mineral is optically positive, and vice versa.*

This statement should be compared with one given on p. 58.

Since the light rays form a cone, the outermost rays have a longer light path through the mineral so that the interference colours rise towards the margin of the figure. The number of colour bands varies with the birefringence of the mineral. A figure given by quartz will normally consist only of isogyres and a grey field perhaps passing to pale yellow at the margin of the figure. Calcite, with its strong birefringence, produces a figure containing a large number of colour bands.

The student can get a good idea of the way in which a uniaxial figure is produced by visualising what would happen if a quartz-wedge were rotated at great speed on the stage, with its thin end at the centre of the field. Parallel light would be used and the polarisers would be crossed. An impression of concentric colour bands would be created. The light vibrations would quite genuinely be radial and tangential. A black cross would be created by the repeated extinction of the light every time the wedge passed the E-W and N-S directions.

In order to determine the optic sign it is only necessary to superimpose an accessory plate upon the vibrations in the figure and to note the colour changes that take place in the four quadrants of the figure. Fig. 15 shows what happens when length-fast gypsum and mica plates are superimposed on the figure of a uniaxial *negative* mineral. In the two quadrants that are in line with the plate the colour value is increased. Thus in these quadrants the radial extraordinary (e′) vibrations are fast, the tangential ordinary (o) vibrations are slow; they correspond in direction with the fast and slow vibrations of the accessory plate, giving rise to the net increase in birefringence. It will be realised that the principles involved in the use of the accessory plate are identical whether one is determining the relative velocities of vibrations in a part of an interference figure, or in the whole of a mineral section viewed under normal conditions between crossed polarisers. A mica plate has relatively little

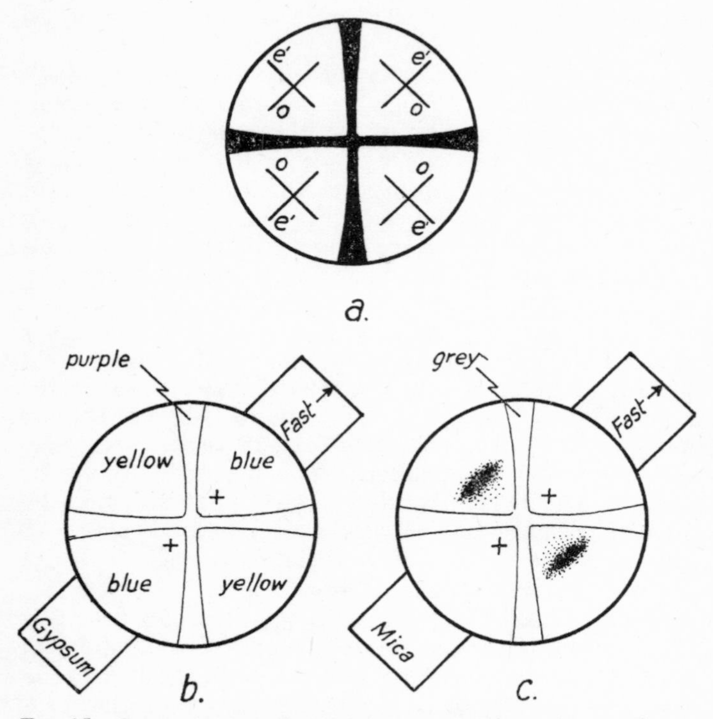

FIG. 15. *Determination of sign from a uniaxial interference figure. For explanation see text. + signs indicate addition of interference colours caused by insertion of accessory plates.*

effect on the colours; but in the two quadrants where subtraction takes place two small dark areas ('dots') develop where there is compensation. The colour produced by the mineral in the figure is here exactly balanced by that of the mica plate. Choice of an appropriate accessory plate depends upon the birefringence of the mineral under examination. If the birefringence is weak and the interference figure is devoid of colour bands, it is best to use a gypsum plate. This is the condition envisaged in fig. 15b.

Strong birefringence and bright colours in the interference figure call for use of a mica plate.

A quartz-wedge may also be used, and in that case attention is focused upon *movement* of the colour bands in the figure. In the quadrants where fast and slow vibrations of both mineral and quartz-wedge are parallel, the total birefringence is increased. The effect is to *increase* the number of bands in the figure, so that as the wedge is inserted the colour bands appear to move in towards the centre of the figure. In the other quadrants, subtraction takes place and the colour bands move outwards.

Of course, with a uniaxial *positive* mineral the effects are precisely opposite to those described.

INTERPRETATION OF BIAXIAL FIGURES

Biaxial interference figures are considerably more difficult to deal with than uniaxial ones. Whereas in a uniaxial figure there are two variable properties—the sign and the number of colour rings controlled by the birefringence—in a biaxial mineral there are three variables. The third is the spacing of the isogyres when they are most widely separated. This is controlled by the angle between the two optic axes (see fig. 14). The acute angle between these axes is referred to as the **optic axial angle** or symbolically as the angle 2V. Measurement of this angle (which varies with composition as do other optical properties) calls for advanced apparatus and techniques. However, the student should be able to distinguish minerals having small (say 0°-30°), moderate (30°-60°) and large (60°-90°) 2V from the appearance of the resulting interference figure. With a small 2V the isogyres scarcely separate and each isogyre is strongly curved. The colour bands are almost circular (Plate I, fig. 1). With a moderate 2V the isogyre separation is quite obvious; but if the figure is well centred, both melatopes (p. 64) remain well within the field of view. If 2V is large the isogyres separate to the margin of the field or beyond, and in this position each isogyre shows only very slight curvature. The colour bands become very elliptical. The most convenient demonstration of varying optic axial angle is provided by cleavage flakes of muscovite, biotite and phlogopite mica.

We have not so far dealt with the **sign of biaxial minerals,** because this is a property that can *only* be determined from an interference figure. The sign depends upon the position of the optic axes with respect to the principal vibrations X and Z. Optic axes always lie within the plane containing X and Z, known as the **optic axial plane.** Furthermore the optic axes are always symmetrically disposed towards X and Z.

In a **biaxial positive** mineral, the vibration direction Z bisects the acute angle between the optic axes. Z is then referred to as the **acute bisectrix** (Bx_a). In a **biaxial negative** mineral, the vibration X is the acute bisectrix. The optic axes lie closer to X than to Z.

The position of the optic axial plane can be determined rapidly from a centred biaxial figure. It passes through the melatopes of the two isogyres when the latter are in their most widely separated positions, as shown in fig. 16a. The acute bisectrix (that is, either X or Z) lies midway between the melatopes, which, of course, is only another way of saying that it is midway between the optic axes, and normal to the plane of the section. A biaxial figure of this kind is referred to as an acute-bisectrix figure.

The vibration directions vary in different parts of the figure because of the convergent nature of the light. In practice, it is only necessary to know the character of the vibrations occurring within the optic axial plane when the figure is in its 45° position. These vibrations, shown in fig. 16a must be compared with the Y vibrations which occur at right angles to this plane. For the purpose of determining the sign it is only necessary to superimpose an accessory plate and to observe whether the colour order at the centre of the figure, between the isogyres, is increased or decreased. One is actually testing the velocity of the **obtuse bisectrix** (Bx_o) vibration relative to that of Y.

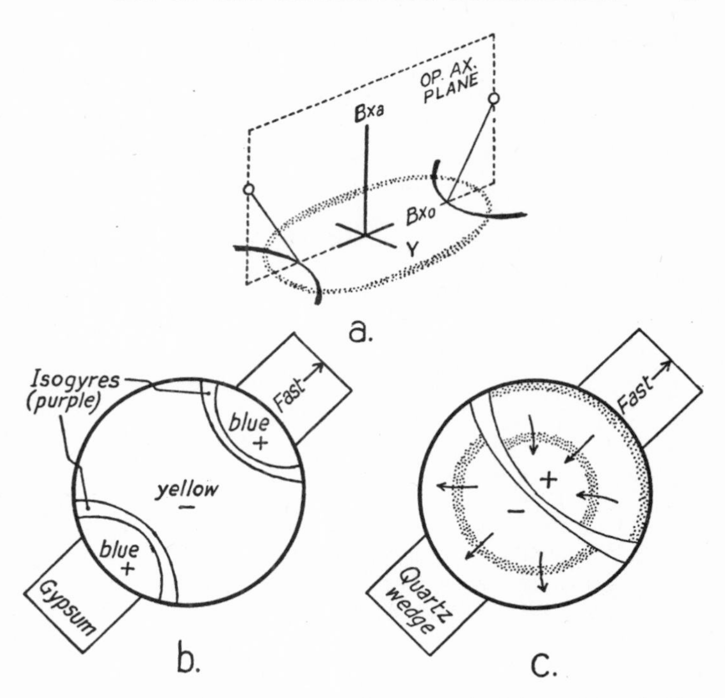

FIG. 16. *Biaxial interference figures. (a) Orientation of Bx_a figure. (b) Determination of sign from Bx_a figure. (c) Determination of sign from optic-axis figure. + and — signs indicate addition or subtraction of birefringence caused by insertion of accessory.*

Fig. 16b shows a typical example, using a length-fast plate, and with the optic axial plane parallel to the plate. The colour order in the centre of the figure is represented as decreasing, so unlike vibrations are therefore superimposed. The obtuse bisectrix vibration is therefore slow compared with Y. Thus Bx_o is Z, Bx_a is X and the mineral is optically negative. (The diagram was prepared using muscovite as an example, so the student can easily reproduce these results in practice.)

THE PRACTICAL USE OF INTERFERENCE FIGURES

Most people find relatively little difficulty in interpretating a perfectly centred uniaxial figure or an acute-bisectrix figure of, say, a cleavage flake of mica. The difficulty arises when searching for such a figure—or, rather, the particular mineral grain in a rock-section which will yield an ideal figure. Fortunately use can be made of figures which are considerably off-centre. This is especially true with uniaxial figures which can generally be recognised by their straight isogyres even when the point of emergence of the optic axis is lying just out of the field of view. The isogyres sweep across the field in a characteristic order as the stage is turned: from east to west, say, followed by movement from south to north, west to east, north to south, and so on. The position of the optic axis can always be ascertained by reference to the partial colour bands in the field of view. With practice an eccentric figure can be made to yield as much information as a centred one. When the degree of eccentricity passes a certain limit, however, there is a danger of confusing a uniaxial figure with a biaxial one. To be safe, therefore, it is best to concentrate attention on figures in which the point of emergence of the optic axis is always in view.

Biaxial interference figures vary considerably according to the orientation of the mineral slice. Probably the safest generalisation is to say that the parts of the isogyres which will appear from time to time in any biaxial figure (as the stage is rotated) cross the field and become inclined to the crosswires somewhere en route. These fragments of isogyres are generally curved.

One special type of eccentric biaxial figure is of particular significance. This is given by a mineral sectioned perpendicular to one of the two optic axes. The appearance of such an **optic axis figure** is shown in fig. 16c. As the stage is turned, the single isogyre rotates in a counter direction. As long as the isogyre is appreciably curved in the 45° position it is possible to deduce the position of the

Bx_a vibration and to determine the optic sign quite as easily as if the whole Bx_a-figure were visible (see fig. 16c).

For distinguishing purposes, an optic-axis figure is often more convenient to use than an acute-bisectrix figure. The optic axes are directions which are sensibly isotropic, so that a suitable choice of mineral-section is easily made. This is an important point for students who only wish to use interference figures as a means of distinguishing uniaxial from biaxial minerals. *The rule then is always to test the grains showing the lowest order interference colours* —that is, approaching most closely to isotropic conditions. There are several cases in which this test helps to differentiate minerals which are closely similar in most other respects. An interesting case is provided by quartz (uniaxial) and cordierite (biaxial). These two minerals are both usually quite colourless and have comparable relief and birefringence. Both minerals tend to form equigranular aggregates with irregular boundaries to the grains. Obviously in a case like this the examination of interference figures can be extremely helpful. Another case is provided by orthoclase (biaxial) and nepheline (uniaxial), especially in coarsely crystalline nepheline-syenites in which the nepheline crystals may lack their characteristic hexagonal shapes.

If interference figures are being examined as part of the complete optical determination of a mineral—involving its optical orientation and sign—then care must be taken to relate the information gleaned from the figure to the cleavage directions and external shape of the section concerned. This is a topic which is really outside the scope of this book, and one example will have to suffice. In the case of barytes shown in fig. 14 all three pinacoidal sections give biaxial interference figures of a kind. The figure from the (001) section is an obtuse bisectrix figure. It can be distinguished from the acute bisectrix figure of the (100) section by the fact that the isogyres separate right outside the field of view when the figure is in the 45° position. Although such a figure is only of limited value and would be useless for determination of sign, for instance, yet it does indicate the position of the optic axial plane and of the vibration Y. It would be noted that the optic axial plane bisects the obtuse angle between the two prismatic cleavages. The most useful figure is, of course, the acute bisectrix figure, and from this the approximate position of the optic axes could be inferred and, if necessary, sketched on to a three-dimensional diagram as in fig. 14. It would be noted that the optic axial plane is perpendicular to the basal cleavage traces of the section: that is, perpendicular to the (001) plane of barytes. The vibration Z would be confirmed as the acute bisectrix, thus giving the sign as positive. Lastly, there is the (010) section. This gives the highest order interference colours, and is therefore the plane containing the vibrations X and Z, or in other words, the optic axial plane. Figures given by the most brightly polarising sections of *any* mineral are almost valueless for determinative purposes, and can actually be misleading. They are, therefore, best ignored.

CHAPTER 7

The Nature of Rock-forming Minerals

Several thousand different mineral species are known. The great majority of these are rarities, either because of their composition or because their ranges of stability in regard to pressure and temperature are extremely limited. Nearly all the material of the earth's outer crust is composed of surprisingly few major elements, listed below in order of abundance.

Element	*Symbol and valency*	*Ionic radius, 10^{-8} cm.*
Oxygen	O^{2-}	1·32
Silicon	Si^{4+}	0·39
Aluminium	Al^{3+}	0·57
Iron—ferrous	Fe^{2+}	0·83
—ferric	Fe^{3+}	0·67
Calcium	Ca^{2+}	1·06
Sodium	Na^{+}	0·98
Potassium	K^{+}	1·33
Magnesium	Mg^{2+}	0·78

Titanium, hydrogen, phosphorous, manganese, sulphur and carbon are also relatively abundant. All other elements are present in very subordinate amounts.

The really common rock-forming minerals are those composed of the elements listed above. Most of these minerals are **silicates,** formed of oxygen and silicon, in combination with certain of the metallic elements. Two factors are dominant in deciding what the combination shall be. In the first place the atoms, which can be pictured as minute spheres, have to have an effective radius which allows them to fit into the structure exactly. Secondly the electric charge on each atom (or ion) has to be such that the sum of all the positive and negative charges in the crystal structure is equal. The structure must be electrically neutral. For these reasons the electric charges or valencies, and the radii are given for the elements listed.

The two-fold criteria of ionic radius and valency limit the number of silicate structures that can be formed. A very brief review may be helpful since it will indicate why many minerals such as augite, diopside and aegirine, say, form one compact group (the pyroxenes) with many shared characteristics of shape and cleavage, while other minerals such as hornblende and tremolite belong to another quite distinct group (the amphiboles). A review of this kind

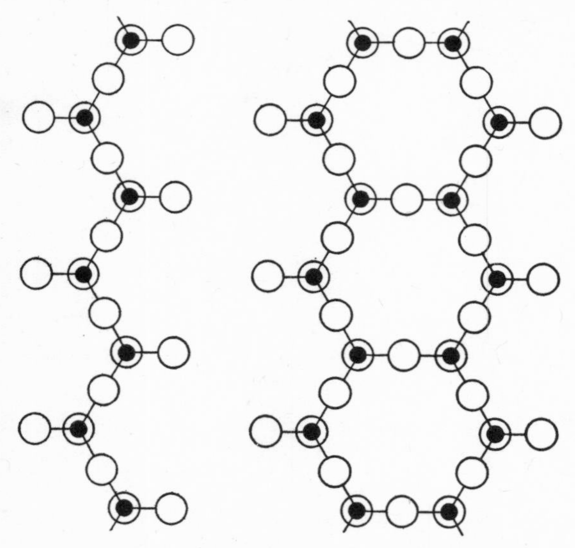

FIG. 17. *Single and double chain silicon-oxygen structures of pyroxenes (left) and amphiboles (right). The chains are elongated parallel to the vertical crystal axes and present the* (100) *aspect.*

brings some sense of order into the lists of compositions which are otherwise meaningless and a burden on the memory.

Silicon and oxygen form a particularly stable combination of ions on account of the small size and large positive charge of silicon compared with the larger size and negative charge of oxygen ions. Four of the latter surround each silicon ion in what is known as a tetrahedral grouping (fig. 17). Obviously an SiO_4-tetrahedron is not electrically stable since there is a surplus negative charge of four units. This charge is neutralised in one of two ways: (1) by linking the tetrahedra through the agency of positive metallic ions; or (2) by sharing oxygen atoms between adjacent tetrahedra. From the descriptions of the main silicate structures, a-f, given below it will be seen that there is a gradual change from 1 to 2.

In each of the structures to be described there is generally 'room' for a certain degree of interchange or substitution of the metallic ions, particularly if these are similar in ionic radius and valency. Such substitution is indicated in a formula by grouping the interchangeable ions in brackets, as in the case of (Fe,Mg). Ionic substitution within an otherwise uniform atomic structure leads to isomorphism between the various members of each group of silicate minerals. Isomorphism (*i.e.* similarity of crystallographic constants and form) is of great importance from the point of view of optical properties of minerals, since the properties vary according to the composition.

The following structural types of silicates can be recognised:

(a) Silicates in which the SiO_4-tetrahedra are linked *via* metallic ions. Each tetrahedron is entire, and there are no 'shared' oxygens. The olivines provide a typical example, in which the metallic ions are Fe^{2+} and Mg^{2+}. If we take a single tetrahedron as a unit of structure, the number of metallic ions required to neutralise the surplus (4−) negative charge is obviously two. The complete formula becomes $(Mg,Fe)_2\,SiO_4$.

FIG. 18. *The sheet or layer structure of mica. Positions of oxygen atoms (large circles) and silicon (small circles) only are shown. Note the pseudo-hexagonal symmetry of the structure.*

(b) Silicates in which the SiO_4-tetrahedra are linked in chains as shown in fig. 17 (*left*). The linkage is achieved by the sharing of oxygens between adjacent tetrahedra. The complete chains are bound together by metal ions not shown in the diagram. In the **Pyroxenes** which have this kind of structure, the metals are principally Ca, Mg and Fe. Isomorphism is achieved in the group mainly by ionic substitution of iron and magnesium. The composition can be calculated approximately by noting the number of silicon and oxygen atoms in the unit of pattern of the diagram and adding Ca, Mg and Fe according to requirements of the valency. In this way we obtain Ca (Mg,Fe) Si_2O_6 as one possible formula.

(c) Silicates with a double-chain ('band') type of structure, in which more of the oxygen atoms are shared between adjacent tetrahedra, as shown in fig. 17 (*right*). **Amphiboles** belong to this category. The formula is slightly complicated by the addition of negatively charged hydroxyl ions $(OH)^-$, which lie in the centre of the rings of silicon and oxygen atoms. Metallic ions present are similar to those in pyroxenes. The student may like to confirm by the procedure outlined above that one formula for amphibole is $Ca_2(Mg,Fe)_5\ Si_8\ O_{22}(OH)_2$.

(d) Silicates with a sheet-like structure as shown in fig. 18 are common and include the micas, chlorites, talc and the clay-minerals. In no other minerals is the relationship between atomic structure and physical properties so marked and so easy to appreciate (p. 96). We may take mica as an example to show how the composition of muscovite (white mica) may be derived. The relative number of silicon and oxygen ions present can be obtained from the unit of pattern enclosed in the rectangle marked on fig. 18. A1 takes the place of one ion in four of the Si atoms, giving $(A1Si_3)O_{10}$. Hydroxyl (OH) is present as in amphiboles, though in approximately twice the quantity. The partial composition $(A1Si_3)O_{10}(OH)_2$ gives a surplus negative charge of 7 units which is neutralised by K and A1 as follows:

$$K\,A1_2(A1Si_3)\,O_{10}(OH)_2$$

The excellent cleavage of mica is developed between layers of silicon and oxygen atoms, since the layers are separated by large potassium atoms which are very weakly bonded to the rest of the structure.

(e) Silicates in which the SiO_4-tetrahedra are linked in all three dimensions by the sharing of oxygens. Aluminium takes the place of some of the silicon in the tetrahedral groupings, so that minerals in this structural category are alumino-silicates. The felspars all belong here. It is difficult, unfortunately, to indicate their structure diagrammatically, and in this brief survey one can only state that Ca^{2+}, Na^+ and K^+ are the other chief constituents of felspars giving (1) $CaAl_2Si_2O_8$, (2) $NaAlSi_3O_8$, and (3) $KAlSi_3O_8$. Isomorphism is particularly important between (1) and (2) and more limited between (2) and (3).

(f) Finally every oxygen of the SiO_4-groupings is shared in a three-dimensional linkage, giving a composition SiO_2, or silica. The geometry of the linkage varies according to temperature and pressure conditions, and in this way several distinct minerals having the composition SiO_2 can occur. By far the most important of these polymorphous minerals is quartz. Since there is no other tetravalent element of small ionic radius that can take the place of silicon in the structure, there is no ionic substitution and quartz is probably the only common mineral which has a really constant composition.

In igneous rocks it is usual to find one or two of the silicates containing iron and magnesium — olivines, pyroxenes, amphiboles and micas—associated with felspars and perhaps with quartz. Most of these minerals are also common in the crystalline metamorphic rocks. In the latter, one may also find certain silicate minerals not mentioned so far, which sometimes achieve rock-forming status. These include *garnets,* the aluminium silicates (*andalusite, sillimanite* and *kyanite*) and *cordierite.*

In any rock, distinction may be made between the predominant **essential minerals** whose presence decides the category of the rocks, and **accessory minerals** which are present in very variable but always small amounts. In a typical granite, for instance, quartz, alkali-felspar and mica rank as essential minerals, while zircon, apatite and magnetite rank as accessories. The list of possible accessory minerals in both igneous and metamorphic rocks is a long one; but the most frequently found accessories are limited. They include *apatite* (the only common phosphate mineral), *sphene* (a titanium mineral), and opaque iron ores of which *magnetite* and *ilmenite* are the chief.

Sedimentary rocks consist of three main types of component: (1) detrital mineral grains derived unaltered from igneous, metamorphic—and, of course, earlier sedimentary—rocks; (2) minutely crystallised secondary minerals produced by weathering and the breakdown of primary constituents; and (3) minerals precipitated from aqueous solution.

Chief amongst the detrital minerals is quartz, which is the predominant component of sands and sandstones. It owes its preservation to its mechanical toughness since it is a very hard mineral which has no cleavage, and also to its chemical stability. There is no weathering product into which quartz can be altered. Felspar ranks second amongst the detrital minerals, and under special conditions, any one of the typically igneous or metamorphic silicates may occur in this way.

Minerals produced by weathering and low-temperature chemical alteration, and especially the so-called **clay minerals,** are too minutely crystalline to be examined with the petrological microscope. Therefore, despite the great importance of clays to the geologist, the engineer and the industrialist, we must perforce ignore their constituents in this book.

The minerals formed by chemical precipitation are mostly salts of common acids such as carbonic, sulphuric and hydrochloric acids. The carbonates, sulphates and chlorides differ fundamentally from the silicates of igneous rocks. Silicates are typically products of high temperature, and in igneous rocks, of actual fusion, which may take place in a 'dry' state with very little or no water present. Carbonates, etc., are formed at atmospheric temperatures and are not only precipitated from aqueous solutions but are relatively easily dissolved in water. The most insoluble of these salts are the carbonates *calcite* $CaCO_3$ and *dolomite* $CaMg\ (CO_3)_2$ which form the main constituents of limestones.

CHAPTER 8

The Ferro-Magnesian Silicates

The dominant ferro-magnesian silicates occurring in rocks belong to the olivine, pyroxene, amphibole, mica and chlorite groups, which are described in that order—based on increasing complexity of atomic structure.

THE OLIVINE GROUP

Composition. The general formula for the several members of this group may be written R_2SiO_4, in which R stands for Fe,″ Mg, and rarely Mn and Zn. In this account attention will be restricted to those members of the group containing Mg and Fe in varying proportions. There is complete isomorphous substitution between the two end-members, **forsterite,** Mg_2SiO_4 and **fayalite,** Fe_2SiO_4. Specific names are used for definite ranges of composition, but opinion is divided as to the delimitation of these ranges. Common **olivine,** sometimes termed chrysolite, may be expressed as $(Mg,Fe)_2SiO_4$ magnesium being in excess, usually large excess, over iron, and averaging about 70 to 80 per cent of the forsterite molecule. **Hortonolite** contains iron in excess of magnesium; while the end members range from forsterite 100 to 90 per cent, and fayalite 10 to 0 per cent of the forsterite molecule respectively.

Occurrence. Common olivine is very widely distributed in basic igneous rocks including basalts, dolerites and gabbros, in which it occurs in association with augite and plagioclase near labradorite in composition. In the lavas (basalts) the olivines occur as well-shaped phenocrysts (first-generation crystals of relatively large size, though they seldom exceed a small fraction of an inch in maximum dimension).

The mode of occurrence of forsterite is normally quite different: it is a metamorphic mineral produced when magnesian limestones (including dolomites) are subject to thermal metamorphism. Forsterite crystals therefore occur in a matrix of crystalline calcite (forsterite-marble).

Fayalite is the only member of the group that is stable in the presence of free silica, and therefore occurs in 'acid' igneous rocks including some of the natural glasses (fayalitic pitchstone), acid lavas (rhyolites) and their hypabyssal equivalents: its associates therefore include quartz and alkali-felspars.

Irrespective of differences in composition, all members of the group are alike in form and general physical (including optical) properties.

Form and Cleavage. The olivines crystallise in the Orthorhombic system, and although chance sections convey little information concerning the solid form of the mineral, six-sided sections as illustrated in Pl. II are distinctive. Generally cleavage, if present at all, is feebly developed, but is more commonly encountered in hortonolite than in common olivine. Cleavage when present is parallel to the side-pinacoid faces, and therefore indicates the direction of the *c*-axis. Strongly developed arcuate fractures are characteristic, and are often emphasised by alteration.

Colour. Olivines are transparent and of a distinctive olive-green colour in natural crystals, but are completely colourless in thin section. Fayalite may show a trace of a faint lemon yellow tint.

Refractive Indices and Birefringence. The properties vary with composition, and in the hands of the expert provide means of finding the exact chemical composition of a specimen by optical means. For all members of the group the refractive indices are high—for common olivine 1·65 to 1·73—causing high relief: strong outlines and rough and pitted surface. The birefringence is approximately 0·035, so that the highest interference colours, shown by (001) sections, are bright second order colours.

Sign and Orientation. 2V is very large (80°-90°) and it may not be possible to determine the sign in consequence. Optic-axis figures obtained from grey-polarising olivine crystals have sensibly straight isogyres. Actually common olivine is optically positive. Sections showing the (010) cleavage to best advantage have straight extinction and are length-slow.

Alteration. Olivine is often recognised in the first place by its alteration to serpentine (p. 99). This alteration proceeds most readily along the fractures noted above, and at the margins of crystals. Veinlets develop with minute fibres of serpentine growing across the veins. With more intense alteration the fresh olivine is reduced to small residual patches showing high relief, in the midst of the serpentine which has a low relief. Finally, serpentinisation may be complete, giving pseudomorphs after the olivine. The serpentine may be almost colourless, green, yellowish-green or brown stained. It is usually charged with opaque iron oxide (magnetite) which is produced as a by-product of the reaction that converts olivine—a silicate containing both Fe and Mg—into serpentine which is a hydrated magnesium silicate essentially free from iron.

Alteration may give rise to secondary talc instead of serpentine. The talc may be recognised by its brilliant interference colours which are similar to those of muscovite, and by its association with olivine. As might be expected from its composition, fayalite when altered by weathering may be represented by strongly ferruginous pseudomorphs—usually deep brown to yellowish brown limonite.

THE PYROXENE GROUP

This is a large group of minerals varying in crystallographic features as well as in composition. Many are important rock-forming silicates occurring widely in igneous and metamorphic rocks.

Composition. The general formula may be written $R^{++}{}_2Si_2O_6$ (or $RSiO_3$ in certain simple cases), where R^{++} is Mg, Fe (divalent) or Ca. While there is perfect isomorphous substitution between the magnesium and ferrous iron, giving the continuously variable series known as the orthopyroxenes, entry of calcium into the structure is limited. The introduction of calcium has an interesting effect on crystal symmetry, as the Ca-bearing pyroxenes are Monoclinic, while those containing Mg and Fe only are Orthorhombic. Actually pure $CaSiO_3$ — the mineral wollastonite — is Triclinic and its atomic structure is different from that of the pyroxenes. As it is linked to the latter by its similarity in composition, it is sometimes referred to as a pyroxenoid.

The common pyroxenes are divided on a crystallographic basis into two series — the clinopyroxenes and orthopyroxenes:

1. **Clinopyroxenes** (Monoclinic).

   Diopside $CaMgSi_2O_6$
   Hedenbergite, $CaFeSi_2O_6$
   Augite, complex composition, approximately $(Ca,Mg,Fe)_2Si_2O_6$ with small amounts of Al, Na, etc.
   Pigeonite, similar to augite, but richer in Mg and Fe and correspondingly poorer in Ca.
   Aegirine, in which the place of $Mg^{++}Fe^{++}$ is taken by $Na^{+}Fe^{+++}$.
   Aegirine-augite, intermediate between aegirine and augite.

2. **Orthopyroxenes** (Orthorhombic).*

   Enstatite, $MgSiO_3$
   Hypersthene, $(Mg,Fe)SiO_3$
   Ferrosilite, $FeSiO_3$

*In dry-melt experiments minerals with the composition of orthopyroxene may occur, but show Monoclinic symmetry: they are therefore conveniently termed clino-enstatite, clino-hypersthene, etc. They are not found occurring naturally.

CLINOPYROXENES

Of the various members of this sub-group, augite is of outstanding importance, as it is by far the best known and most wide-spread. **Augite** occurs as euhedral crystals, often of perfect form (Pl. III), in certain basic lavas, including basalts and the significantly named augitites. It also occurs in intrusive rocks such as dolerites, gabbros and pyroxenites.

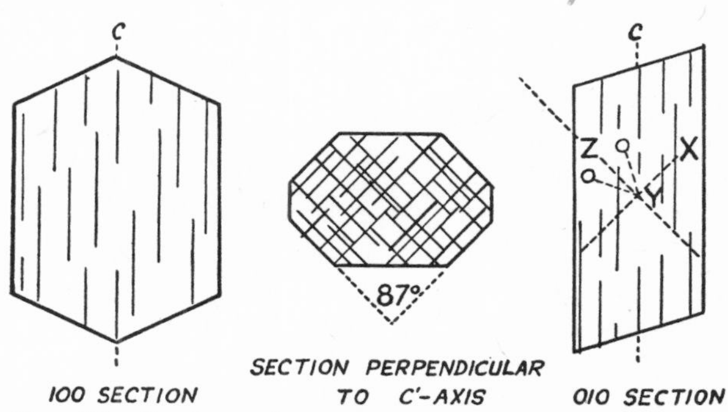

FIG. 19. *Sections of augite showing cleavages and the optical orientation. The angle between Z and c is the extinction angle.*

Form and Cleavage. The characteristic outlines of pinacoidal sections of augite are illustrated in fig. 19, and these may readily be distinguished among the sections occurring in slides of basalts, in which the larger crystals are embedded in a matrix of very small grains or microlites, including second-generation augites.

Colour. Augite crystals are invariably black in hand specimens, but are light brown or greyish brown in thin section, while those near to diopside in composition may be sensibly colourless. As regards colour, two points are noteworthy. One of the rarer elements in some augites is titanium; and so-called titanaugite is distinctively tinted mauve in thin section. Again, the incoming of Na appears to result in a change of colour to bright green, in aegirine-augite. Both of these varieties show a degree of pleochroism; but as a rule common augite is non-pleochroic.

Specific Optical Properties. The refractive indices vary with composition; but lie within the limits 1·68 to 1·74, giving moderately high relief. The birefringence is approximately 0·025, so that the highest interference colours, shown by (010) sections are second order blues.

The optical orientation is that found in many Monoclinic minerals, with the optic axial plane parallel to (010), and the optic normal (Y) coincident with the crystal axis, b. 2V is moderate, generally about 55°, and the sign is optically positive (fig. 19).

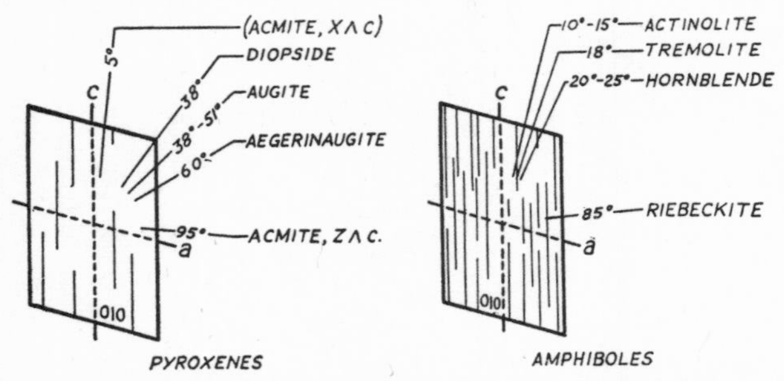

FIG. 20. *Extinction angles of some Monoclinic pyroxenes and amphiboles. Angles measured between the slow vibration Z and the c-axis.*

As regards extinction, it follows from the above that side-pinacoidal sections show oblique extinction at the maximum possible angle. Actually careful measurement (preferably on the Universal Stage) of the maximum angle indicates the precise composition of the specimen under observation, as shown in fig. 20. By contrast, the (100) section shows *straight* extinction. Faces in zone with (100) and (010) may vary in extinction angle, therefore, between, say, 51° and 0°. A section perpendicular to the prismatic cleavages and, therefore, to the *c*-axis, extinguishes in the position shown in fig. 19.

Special Features. Twinning, both simple and multiple, is common, in both cases the twin plane and composition plane being (100). In natural crystals such twinning may be readily recognised by the occurrence of a re-entrant notch at the top of the crystal, but in thin section it is strikingly demonstrated by differences in the colour of the different parts of the twin. Even in plane polarised light a slight difference of the absorption betrays the twinning; but with crossed polarisers striking differences in the interference colours make it much more obvious.

Very slight differences in composition are reflected in appreciable differences in birefringence; so that zoned augites are not uncommon (Pl. III). In many of these, lighter and darker zones alternate, suggesting a rhythmic change of composition during crystal growth; but in some cases a more drastic change is indicated by, for example, a narrow, regular bright green external zone surrounding a light greyish brown core. This is an external zone of aegirine-augite on a core of common augite, and the corresponding change in the composition of the melt is obvious. 'Zoning' of yet another type gives rise to '*hourglass structure*' in augite (Pl. III). In crystals showing this phenomenon a single crystal grain shows two contrasting interference colours affecting different parts in a manner recalling the shape of the ancient hour-glass.

In measuring extinction angles of augite it will frequently be found that complete extinction is not achieved: near the position of mean extinction the interference colour is seen to change from a bluish to a brownish colour. This is due to *dispersion*—the non-coincidence of the extinction positions for the several components of white light. When blue is extinguished, a little red comes through and conversely.

Finally, augites may show *schiller structure*. Under the microscope this is seen as very minute hair-like opaque inclusions lying with rigidly parallel orientation in certain planes, and representing material that was exsolved (thrown out of solution) with falling temperature. In other cases the minute inclusions are planar, and in coarse-grained rocks may be sufficiently thick to produce an appreciable effect between crossed polarisers, when it can be proved that the inclusions consist of lamellæ of orthopyroxene embedded in the augite.

Diopside is green in hand-specimens, and very light green to colourless in thin slice, otherwise its optical properties are like those of augite. Diopsidic pyroxenes do not occur in basic igneous rocks, though they are not uncommon in 'acid' lavas; but the mineral is typically metamorphic, and, as might be inferred from its composition, is produced by metamorphism of Mg-bearing limestones. It occurs typically as very light green to colourless crystal grains embedded in crystalline calcite (marble) sometimes associated with forsterite and wollastonite.

Aegirine is much rarer than augite, it is restricted to igneous rocks rich in sodium, and is associated with alkali felspars and felspathoids such as nepheline. As noted above, its distinctive features are bright (almost emerald) green colour, pleochroism to more yellowish tints, and small maximum angle of extinction. The crystals are often notably elongated, and even for pyroxenes, the cleavages appear to be particularly perfect.

Aegirine-augite. The diagnostic characters of this mineral are adequately covered by the statement that it has the colours and pleochroism similar to aegirine with the extinction of augite.

Pigeonite has been recognised increasingly of recent years to be an important mineral in many basalts. Crystals cannot be distinguished from common augite unless the interference figures are examined. Pigeonite has a very small 2V, and an optic plane perpendicular to that of augite.

For the sake of completeness a note on **wollastonite** is included, though it is not a mineral for a beginner. If, however, in a section identified on other criteria as a lime-silicate rock (metamorphosed limestone) a colourless mineral is encountered with the cleavage characteristics of a pyroxene, but a definite bladed to fibrous habit, it may be suspected of being wollastonite. The blades are sometimes length-fast, sometimes length-slow, while the indices are appreciably less than augite, and the birefringence considerably weaker (about 0·15).

ORTHOPYROXENES

In the continuously variable series of orthopyroxenes arbitrarily chosen limits define a number of named mineral species, of which the more important are named in the list above. Ferrosilite is very rare indeed and is therefore excluded from further consideration here.

Crystallographic Features. Euhedral crystals of orthopyroxenes are very rare in mineral collections, and need not be considered here. In thin sections small phenocrysts embedded in a fine-grained groundmass are common, especially in andesitic lavas. In general the shapes of these crystals in thin section (Pl. III) are closely similar to those of augite; though the eight-sided basal sections tend to be more regularly octagonal on account of the approximately equal development of the front- and side-pinacoids. As in augite, the cleavage is prismatic, and is equally good.

Colour. **Enstatite** is colourless in thin section, while **hypersthene** is nearly so. In a relatively thick section, however, the colouring of hypersthene is distinctive: it is pleochroic from light bluish green to light pink, rather like that shown by some andalusites. It is customary to apply the name hypersthene to those orthopyroxenes which show colour and pleochroism in thin section, and the name enstatite to those which do not. As the amount of light transmitted depends upon the thickness of the slide, this is obviously only a rough and ready means of distinction. More accurate distinction involves the study of the optical orientation of the mineral—its sign and optic axial angle.

Refringence and Birefringence. The refractive indices increase in the isomorphous series from enstatite to hypersthene. For the latter $n\alpha$ is approximately 1·69, and $n\gamma$ 1·70.

The relief is therefore similar to that of augite. Birefringence, approximately 0·01, is considerably less than that of augite, and the interference colours are low in the first order of Newton's Scale: grey is characteristic. This feature, together with the straight extinction of all principal sections, helps to distinguish orthopyroxenes from clinopyroxenes. It is very important, however, to realise that certain sections of augite, in the zone of faces parallel to the b-axis must be perpendicular to an optic axis, which is a direction of single refraction. The interference colour shown by these sections is grey. Further, these sections show straight extinction. Thus we have here *sections of augite showing the features upon which the recognition of enstatite (or hypersthene) is based.*

Optical Orientation and Sign. The optic axial plane is parallel to the side-pinacoid (010). Because of Orthorhombic symmetry, the three vibration directions X, Y and Z correspond in position with the three crystal-axes. Straight (or symmetrical) extinction is the rule, but oblique extinction does not necessarily mean that the section is a Monoclinic pyroxene: it indicates merely that the section is somewhat oblique to the crystal axes.

Relation to the Olivine Group. It is desirable to stress the chemical similarity between corresponding members of the two groups: between forsterite, Mg_2SiO_4 and enstatite, $Mg_2Si_2O_6$; and between common olivine, $(Mg,Fe)_2SiO_4$ and hypersthene, $(Mg,Fe)_2Si_2O_6$. When enstatite is heated it undergoes incongruent melting forming forsterite and siliceous liquid. Conversely when a melt containing the substance of enstatite cools, the first crystals to form are forsterites. Subsequently reaction between these early forsterite crystals and the melt may lead to the conversion of the former into enstatite, either marginally or wholly, depending upon the time factor. A similar relationship holds for hypersthene and common olivine. It is reasonable to expect to find these pairs of related minerals in close association in rocks; and it will be frequently observed that orthopyroxene envelops olivine or forms a thin mantle (a 'reaction rim') around it which makes it quite clear that the two minerals crystallised successively.

Relation to Pigeonite. Pigeonite (*Monoclinic*) is restricted to basic rocks which were quickly cooled from a molten condition. In conditions involving slow cooling, pigeonite proves to be unstable and breaks down into two components — an orthopyroxene, the dominant partner, and clinopyroxene, the subordinate partner which forms regularly orientated lamellæ with the former. The clinopyroxene lamellæ under favourable conditions may be recognised by their bright interference colours, which contrast with the grey of the host mineral.

THE AMPHIBOLE GROUP

Composition. A close analogy exists between pyroxenes and amphiboles in regard to composition. The main metallic cations are the same in both groups, and each pyroxene has its corresponding amphibole; but there is a fundamental difference, as the amphiboles contain hydroxyl $(OH)^-$.

The more important members of the group are:

1. **Monoclinic amphiboles.**

   Tremolite, $Ca_2Mg_5Si_8O_{22}(OH)_2$

   Actinolite, $Ca_2(Mg,Fe)_5Si_8O_{22}(OH)_2$

   Hornblende similar, but with some Al and Na

   Barkevikite, like hornblende but with more Na, and Ti (suggesting analogy with titanaugite)

   Riebeckite, derived, as it were, from the formula of tremolite by substituting $Na'Fe'''$ for $Ca''Mg''$.

   Glaucophane, similarly related to tremolite in which $Na'Al'''$ is substituted for $Ca''Mg''$.

   Cummingtonite, $(Mg,Fe)_7Si_8O_{22}(OH)_2$

2. **Orthorhombic amphibole.**

   Anthophyllite, $(Mg,Fe)_7Si_8O_{22}(OH)_2$

Of this rather formidable list the elementary student will probably encounter only hornblende, which in this group occupies the position of augite among the pyroxenes.

Monoclinic Amphiboles

Hornblende is by far the most widely distributed of all the amphiboles and appears to be stable under very variable conditions, as it occurs in a variety of igneous (particularly Intermediate) rocks and in several types of metamorphic rocks too.

Form and Cleavage. The shapes of pinacoidal sections of hornblende and the appearance of the prismatic cleavages have been described above at some length (p. 28 and fig. 11). The crossing angle of the prismatic cleavage traces at 56° and 124° in transverse sections is as diagnostic of the amphiboles as are 87° and 93° for the pyroxenes. Basal sections are six-sided, with the unit vertical prism rather better developed than the side-pinacoid (Pl. IV).

Twinning, usually simple, but sometimes repeated, is common, the twin-plane and composition-plane being (100) as in augite.

Colour and Pleochroism. Hornblende crystals are black in the hand-specimen, but some shade of green in thin slice. They are strongly pleochroic in shades of green to yellow.

Refringence and Birefringence. The indices of hornblende vary widely with composition and it is pointless to quote figures; but the relief is moderately high (though this feature is of no diagnostic value in the case of hornblende). The birefringence is strong — about 0·02 — and the interference colours are therefore much the same as those of augite; but they are less pure on account of the masking effect of the body colour — the interference colours are viewed as it were through a coloured filter.

Optical Orientation, Sign and Extinction. Optical orientation is the same as for augite: the optic axial plane lying parallel to (010), and the optic normal (Y) coinciding with the crystallographic b-axis. Therefore in a basal section the optic axial plane bisects the *obtuse* cleavage angle, while 'Y' similarly bisects the acute angle.

Extinction is symmetrical in a basal section; it is straight in a front-pinacoidal section, and oblique (at the maximum angle) in a section parallel to the side-pinacoid. This maximum angle is about 25°.

Distribution and Relationship to Augite. Euhedral hornblende crystals occur in certain Intermediate lavas and dyke rocks, notably in hornblende andesites and hornblende lamprophyres. In coarse-grained rocks it forms shapeless grains which, in extreme cases, may make up almost the whole of the rock: this is the case with some significantly named hornblendites; though usually the hornblende is associated with plagioclase near andesine in composition. In deep-seated igneous rocks hornblende not infrequently mantles augite, and sometimes replaces the latter in a highly irregular fashion.

Tremolite and actinolite may conveniently be considered together as the one grades into the other by the atomic substitution of Fe^{++} for some of the Mg^{++} in tremolite. Tremolite is quite white, and therefore colourless in thin section, and normally acicular to fibrous in habit, and occurs typically in metamorphosed limestones, especially magnesian limestones. It is a product of regional, rather than thermal, metamorphism.

The incoming of iron gives a green colour to actinolite, which is of prismatic to acicular habit, occurring notably in highly lustrous rich green prisms, in material from one well-known Alpine locality, embedded in talc. In thin section actinolite is rather like hornblende in general appearance, though the colour is lighter and the pleochroism therefore less pronounced (Pl. IV). A notable crystallographic difference is the virtual elimination of the side-pinacoid, so that cross sections of actinolite tend to be four- rather than six-sided. The extinction angles for these and the other Monoclinic amphiboles are shown in diagram form for easy reference in Fig. 20. These are maximum angles, measured in (010) sections between the slow vibration direction (Z) and the *c*-axis, indicated by the cleavage traces.

Riebeckite among the amphiboles corresponds to aegirine among the pyroxenes. Like the latter it is restricted to soda-rich igneous rocks, and is distinctive by reason of its very strong colour and pleochroism. At its best, riebeckite shows a dark indigo blue colour for some sections in a slide, a dull green for others and brownish green for the remainder, depending on which of the three vibration directions, X, Y or Z, is reaching the eye.

Glaucophane (Pl. IV) is known only in certain rare glaucophane-schists and at its best displays a very striking pleochroism scheme, involving Prussian blue, light purple and grey. It has been necessary to repeat the phrase 'at its best' because atomic substitution between the components of riebeckite and those of glaucophane readily takes place and gives minerals with intermediate characters. Exact measurements of the refractive indices and of the extinction angles are necessary in many cases before identification is certain.

Barkevikite. This Monoclinic amphibole differs from the two preceding by being brown in thin section and quite violently pleochroic. Some varieties show a deep, rich red brown in the position of maximum absorption, through lighter brown to clear yellow in the position of minimum absorption. Other types, notably rich in titanium, appear deep brownish black in one setting of the stage. From fig. 20 it will be noted that the *maximum* extinction angle for barkevikite is 10°. There is therefore some little danger of misidentifying an elongated prism of barkevikite as biotite: the colours and pleochroism may be identical; but the difficulty will not arise if the amphibole is adequately represented in the slide by variously orientated sections, as some of them are bound to show the characteristic cleavages crossing at the usual amphibole angle.

Barkevikite is commonly associated with titanaugite: not infrequently the two show the same kind of relationship as that described above between common augite and hornblende.

Orthorhombic Amphiboles

Anthophyllite is a 'sack name' applied to a suite of minerals corresponding in composition to the ortho-pyroxenes. Some specimens are Mg-rich, others Fe-rich—all are restricted to metamorphic rocks. When examined in thin section it is immediately evident that one is dealing with an amphibole, which typically resembles a light brown actinolite. As the mineral is Orthorhombic, however, all principal sections yield straight extinction. The prisms are length-slow, with Z parallel to the c-axis.

THE MICA GROUP

Mica is a well-known mineral even to those with no geological knowledge, largely because of its economic value as an insulator in electrical industry. As a rock-forming silicate it is widely distributed: one or other of the varieties of mica spans the whole range of composition of the igneous rocks, from the most acid to the most basic; while micas are abundant in many rocks of metamorphic type also.

Composition. All micas are hydroxyl-bearing alumino-silicates; most contain potassium as an important constituent, and, as distinct from the pyroxenes and amphiboles, none contains calcium. The chief micas fall into two descriptive sub-groups, the light and dark micas respectively, the terms referring to their appearance in the mass. The dark micas contain iron and magnesium.

Muscovite, $KAl_2(AlSi_3)O_{10}(OH)_2$
Phlogopite, the bronze mica, $KMg_3(AlSi_3)O_{10}(OH)_2$
Biotite, black mica, $K(Fe,Mg)_3(AlSi_3)O_{10}(OH)_2$.

Of the light micas, **muscovite** is by far the most abundant and for general purposes the two terms are synonymous, for the different varieties of white mica are indistinguishable in thin slice. Certain light micas contain lithium as an essential element. The best known of these is **lepidolite,** which occurs in complex pegmatites. It is distinctively coloured a pleasing shade of light mauve in the hand-specimen.

Phlogopite occurs somewhat rarely in certain types of very basic lava, in which it is often associated with the felspathoid, leucite. It also occurs in impure metamorphosed limestones—clayey magnesian limestone would provide the necessary raw materials. A phlogopite-marble formed in this way is a striking-looking rock, consisting largely of small lustrous bronze phlogopite crystals embedded in a coarsely crystalline matrix of white calcite.

Biotite is much commoner, and is characteristic of, and obvious in, mica-schists, to which it imparts the property of foliation. It occurs also in igneous rocks of almost any degree of basicity, but most abundantly in granites, in so many specimens of which the small, black, highly lustrous crystals may easily be distinguished.

Relationship Between Internal Structure and Physical Properties. In no other type of silicate is the influence of the internal arrangement of the atoms upon the physical properties so obvious as in the case of the micas.

It is obvious from fig. 18 that the mica structure is six-sided in plan, and it is therefore not surprising that euhedral crystals appear to be Hexagonal. Actually the atomic layers are stacked one upon the other with a slight regular offsetting, so that the resultant crystal form is Monoclinic (pseudo-Hexagonal).

The perfect cleavage has been already described (p. 27 and Pl. V). The cleavage planes are those which contain the K-atoms.

When we come to the optical properties we find that the relative ease of vibration of the light is controlled by the atomic structure: the 'easiest' direction of vibration, corresponding to 'X' and the smallest refractive index, is perpendicular to the atomic sheets. The other two, Y and Z, take place *along* the cleavages. As might be expected from the symmetry of the structure, there is little difference between the velocities of these two vibrations, and therefore little difference between the beta and gamma indices. Similarly, in the coloured micas there is no appreciable difference between the absorption tints for Y and Z: both are strongly absorbed and appear equally dark; but the X vibration is much less absorbed and appears correspondingly much lighter. The resulting pleochroism scheme for phlogopite and biotite is shown in the table below.

	X (α)	Y (β)	Z (γ
	(normal to cleavage)	(parallel to cleavage)	
Muscovite	R.I. 1·55	1·58	1·59
	Colour: nil	nil	nil
	2V: 50		
Phlogopite	R.I. 1·535	1·564	1·565
	Colour: nil	reddish brown	golden yellow
	2V: 10		
	Absorption: X<Y<Z		
Biotite	R.I. 1·56	1·60	1·60
	Colour: yellow	dark brown	dark brown
	2V: a few degrees only		
	Absorption: X<Y=Z		

Optical Orientation, Sign and Extinction. The optic axial plane occupies one of two positions: in muscovite it is perpendicular to (010); but in biotite and phlogopite it is parallel to that plane. In all micas, as stated above, X is perpendicular to the basal pinacoid and therefore to the cleavage traces. Further, this is invariably the acute bisectrix, so that all micas are optically negative. This orientation makes mica by far the most valuable material for the study of interference figures, for starting with a cleavage plate a millimetre or less thick, it is possible to cleave the plate progressively thinner and hence to study the effect of thickness on the interference figure. Further, given a representative collection of specimens of mica, small fragments of cleavage flakes of approximately equal thickness serve excellently to illustrate the appearance of interference figures with different optic axial angles. Muscovite shows an ideal biaxial figure, perfectly centred and with the isogyres just off the edge of the field when in the 45° position. With biotite, on the other hand, 2V is so small that the separation of the isogyres on rotating the stage is by no means obvious — the mineral is *pseudo-uniaxial*. Phlogopite is intermediate in this respect, but closer to biotite than to muscovite: it is obviously biaxial, though 2V is small. The limit is reached in lepidotite, which appears to be perfectly uniaxial.

Special Points. Biotite has a wide variation in chemical composition, which results in perceptible differences in the optical details. Thus in some specimens the absorption colour for Z may be brownish black, in others, chestnut brown, or deep red-brown, while in some specimens it is green. For all of these, the colour for the X vibration is light yellow, usually light straw. A frequent cause of confusion to the beginner is the extraordinary difference of appearance between the basal and vertical sections of all micas. With muscovite the basal sections, with a birefringence of about 0·01 (see above table), give an interference colour in the first order grey; and these stand in stark contrast to the vertical sections with their vivid blue, green, yellow and red. In shape also the two sections are markedly dissimilar; while the difference as regards cleavage is self-evident. It is not to be wondered at that these two sections are often misidentified as two distinct minerals; and even more often, the grey-polarising basal sections of white mica are overlooked altogether when, as is usually the case, they are associated with grey-polarising quartz and felspar.

The same difficulty is experienced with biotite. It is difficult to believe that the shapeless, uncleaved, dark-coloured, isotropic basal sections belong to the same mineral as provides the more or less rectangular sections (though with ragged ends), with perfect cleavage, strong pleochroism and vivid interference colours. It is a good plan to look very carefully at the margins of the basal sections when the sheeted structure may be inferred, as some thin overlapping plates project beyond the boundaries.

Zoning may be conspicuous in hand-specimens of mica, but is less frequently seen under the microscope. Rarely, however, the biotites in mica-lamprophyres — dyke rocks associated with granites — provide beautiful examples of zoned crystals.

Again, twinning in a cleavage plate of phlogopite, say an eighth of an inch thick, may be strikingly displayed when viewed obliquely in daylight, as one part of the twin appears deep bronze, while the other part is deep yellow. Definite twinning, visible in thin section, is very rarely encountered, as there is no appreciable difference in absorption between the Y and Z vibrations for this small thickness of material.

Relationship Between White Mica and Felspar. If the formulæ of orthoclase (or microcline) and muscovite are compared, it will be seen that the elements present are the same, except that the latter contains hydroxyl, which is absent from the former. Actually white mica, identical in composition with muscovite, is produced by the hydrothermal alteration of potassic felspar. Usually the mica occurs in the form of minute flakes; but occasionally they are large enough to show definite optical reactions characteristic of muscovite. It is customary to call this *secondary* white mica with the flaky habit by a different name—sericite.

THE CHLORITE-SERPENTINE GROUP

The minerals commonly known as chlorite and serpentine belong to one group of hydroxyl-bearing silicates, which, with one exception, are structurally related to the micas in the sense that they are built up of hexagonal sheets of linked SiO_4 tetrahedra. This accounts for the close superficial resemblance between the micas and the chlorites.

Composition. One end-member of the group is 'serpentine' — the mineral **antigorite,** with the formula $Mg_3Si_2O_5(OH)_4$. Noteworthy features are the large amount of hydroxyl present, and the fact that the mineral is non-aluminous. With the incoming of Al^{+++}, antigorite grades into chlorite of the **penninite** variety, and this in turn into **clinochlorite** (or clinochlore). The Al, together with a little Fe, is substituted for some of the original Mg and Si.

Occurrence. Both chlorite and serpentine are widely distributed in rocks. They are normally secondary minerals; but primary chlorite is sometimes formed very late in the crystallisation sequence. The rock serpentine, represented by the Cornish 'serpentines' in the Lizard District, consists essentially of two minerals — antigorite and chrysotile. These two are identical in composition, but of very different habits: antigorite, like the chlorites, is platy; but chrysotile consists of bundles of parallel fibres, recalling the fibrous amphiboles. Indeed, when the fibres are sufficiently well developed, the mineral is exploited as asbestos. These two minerals commonly replace olivine, as might be expected from their composition. Orthorhombic pyroxene — non-aluminous—alters into a variety of serpentine in which the fibres are arranged parallel to the *c*-axis of the original enstatite or hypersthene. This form of serpentine is known as bastite; and bastite pseudomorphs after orthopyroxenes are familiar features of ancient andesitic lavas.

Chlorites, on the other hand, being aluminous, are alteration products of augite, hornblende, the micas (Pl. V) and garnets, of the appropriate composition. As common augite, hornblende and some garnets contain calcium, alteration into chlorite involves the formation of a complementary alteration product also. This is usually epidote, which forms small rosettes or scattered grains, embedded in the chlorite; but calcite may occur in addition to, or instead of, the epidote.

Quite apart from these types of occurrence as a secondary mineral, chlorite is important in another field.

One of the zones recognised in areas of regional metamorphism is the chlorite-zone, within which chlorite-schist is an important rock-type.

Form and Other Physical Features. The fact that chlorites, including antigorite, are Monoclinic, pseudo-Hexagonal, has been established by X-ray analysis; it can seldom be put to the test of direct observation, as these minerals usually occur as dark green scales or crypto-crystalline felt-like, radial or spherulitic aggregates (Pl. V). Larger crystals do occur, however, in chlorite-schists, and then the resemblance to dark green biotite is very close, especially as regards lustre and cleavage.

Optical Features. Both serpentine and chlorite are typically green in thin section, though the colour may be very pale, often rather yellowish, while some varieties are without appreciable colour. In penninite a light green to yellowish green pleochroism occurs. Refractive indices are generally low, of the order of 1·55 to 1·60. Birefringence is invariably weak, often very weak, yielding dull grey interference colours. Penninite is distinctive in this respect as between crossed polarisers it displays attractive anomalous interference colours, typically a rich indigo blue, but sometimes purple or rich brown. The only other mineral which shows this 'ultra-blue' interference colour is zoisite; but there is not the slightest chance of mistaking the one for the other, as they differ in every other respect. These colour effects result from a combination of particularly weak birefringence (maximum 0·003) with very strong dispersion.

Clinochlorite is very similar to penninite, but without the optical anomalies, while lamellar twinning is characteristic.

Summarily the diagnostic features of chlorites are green colour, slight pleochroism, perfect basal cleavage, straight extinction, and weak birefringence which is sometimes anomalous.

There are other chlorites — Winchell lists thirteen varieties—but they are less common than those described above.

CHAPTER 9

The Felsic Minerals

In this chapter are described the principal light coloured rock-forming minerals, quartz, felspars, felspathoids and the zeolites. The first two are quantitatively very important in rocks, the felspathoids are correspondingly rare, while the zeolites are included here on account of their close chemical affinity with the felspars and felspathoids.

THE SILICA MINERALS

There are several different crystalline forms of silica (SiO_2), of which quartz is by far the most important, being perhaps the commonest mineral on the earth's surface.

Quartz. Two forms are known, one stable above, and the other below, 573 °C. The high temperature form is beta-quartz, but with falling temperature this inverts into alpha-quartz. The two minerals have much in common, though there are crystallographic differences, beta-quartz being genuinely Hexagonal (though in a class without any planes of symmetry), while alpha-quartz belongs to the corresponding class in the Trigonal system. All quartz phenocrysts, such as occur in rhyolites and quartz-porphyries, retain the form of beta-quartz, though of course the substance of these crystals is alpha-quartz at atmospheric temperatures. The properties described below are those of alpha-quartz—the quartz occurring in quartz veins.

Though quartz provides singularly perfect Trigonal crystals for mineral collections, these are obtained either from pegmatites or from cavities in mineral veins. In ordinary rock-sections quartz crystals do not occur; but irregularly shaped grains, with highly indented or sutured margins, are to be expected on account of the relatively low temperature at which the quartz crystallised. The exceptions are the phenocrysts occurring in quickly cooled igneous rocks. These have retained the form of stumpy hexagonal bipyramids, about as tall as they are broad. Consequently vertical sections tend to be four-sided—often nearly square, while basal sections are naturally six-sided. These crystals are often penetrated by corrosion channels produced when the surrounding melt was still very hot (Pl. VI). Although the colour may vary in hand-specimens from almost black through shades of brown to yellow (citrine) and violet (amethyst), in thin slice quartz is invariably colourless and perfectly transparent. There is no cleavage, though arcuate or irregular fractures are common.

Both refractive indices are slightly above that of the embedding medium ($n\omega$ 1·544 and $n\varepsilon$ 1·553), but the difference is so small that even with the light diaphragmed down, in plane polarised light a grain of quartz is practically invisible.

The birefringence, 0·009, is so constant that quartz is used as a standard of reference in optical mineralogy. The thickness of a section may be estimated by the interference colour of quartz, which is faintly tinged with yellow in a section of normal thickness, say 30 microns.

Quartz is uniaxial, positive, though it must be admitted that the sign is less readily determined than is the case with most other uniaxial minerals, as it is virtually impossible to fix the position of the *c*-axis; but isotropic basal sections are easily recognised, and these yield an interference figure under optimum lighting conditions.

Special Features. The properties of quartz in thin section are rather negative in character—negligible surface relief, weak birefringence, no colour, no twinning,* no cleavage

* Twinned quartzes are actually quite common, but with rare exceptions are completely interpenetrant, and show no evidence of twinning in thin section. The twinning may be made apparent, however, by 'sandwiching' a quarter-inch basal section of quartz between two crossed polaroid plates, and viewing it in strong daylight.

and usually no distinctive outline; but actually the identification of quartz is based upon these items of negative evidence. The student may at first find some difficulty in distinguishing between quartz and untwinned felspar; but careful examination will usually show that the latter is cleaved, and altered to some extent, which rules out quartz. *Quartz is invariably fresh and free from alteration of any kind.*

Minute inclusions are common in quartz and may be solid, liquid or gas. Among the mineral inclusions, minute rutiles are noteworthy, as described on p. 130. Liquid and gaseous inclusions may be so abundant as to render the quartz milky in the mass; but usually they are less abundant, and tend to lie along curved lines representing arcuate fractures. Some of the fluid-filled cavities are found to contain mobile gas bubbles (much more heavily outlined than the liquid inclusions) and these sink—apparently upwards! —on rotating the stage. Again, the gas-filled cavities are often bounded by plane surfaces parallel to possible crystal faces of the host mineral: these are so-called *negative crystals*. The appreciation of these features calls for critical focusing and optimum illumination.

Quartz responds readily to mechanical deformation under metamorphic conditions, and even in many normal igneous rocks shows the effects of pressure by the development of 'strain shadows'—shadowy lamellæ visible when the crystal is near the position of extinction.

The high-temperature forms of silica — tridymite and cristobalite—lie outside the scope of this book; but the massive, cryptocrystalline varieties of silica, collectively termed chalcedonite (or popularly chalcedony), merit attention.

Chalcedony (chalcedonite) is a very finely crystalline aggregate of minute granules or fibres, the individual units of which show optical properties indistinguishable from those of quartz, though the mineral is sometimes regarded as a distinct species. Flint and chert consist largely of chalcedony, while in agate the latter is associated with amorphous (and therefore isotropic) opal. It occurs also as vesicle-infillings in lavas [Pl. VI (No. 5)].

Optically chalcedony is distinctive by reason of its cryptocrystalline habit, and a well-marked tendency to form radial or spherulitic aggregates. In such aggregates any fibres lying parallel to either crosswire will be extinguished between crossed polarisers, while those lying in the 45° positions will show their maximum interference colour. Thus a distinctive black cross is formed, which remains stationary when the stage is rotated, as successive fibres are brought into the extinction position.

FELSPAR GROUP

Composition. The felspars are aluminosilicates of sodium, potassium, calcium and rarely barium. The composition of any felspar can be stated in terms of three end-member 'molecules', all of which occur as natural minerals*:

Mineral	*Symbol*	*Composition*
Orthoclase	Or	$KAlSi_3O_8$
Albite	Ab	$NaAlSi_3O_8$
Anorthite	An	$CaAl_2Si_2O_8$

Most natural felspars, however, consist of a proportion of all three components because of the widespread ionic substitution that occurs in the group. Such substitution is complete between albite and anorthite and gives rise to the most important isomorphous series of minerals found in nature, the **Plagioclase Series.** Substitution is limited between orthoclase and albite and virtually non-existent between orthoclase and anorthite, because of the large size of the potassium ion compared with sodium and calcium. Orthoclase and albite are sometimes grouped together as the alkali-felspars to distinguish them from plagioclase. The real mineralogical distinction, however, is between

* The barium felspar, celsian, $BaAl_2Si_2O_8$, on account of its rarity is omitted from this account.

orthoclase which is Monoclinic, and all of the plagioclase felspars and microcline which are Triclinic.

The composition of a particular felspar can be stated in terms of the molecular symbols as in the following examples: $Or_{85}Ab_{15}$ (a sodic orthoclase); $Or_3Ab_{75}An_{22}$ (albite-rich plagioclase) and $Ab_{40}An_{60}$ (somewhat calcic plagioclase). The sum of the molecular percentages must, of course, always equal 100. Since the amount of the Ab-molecule is small in most potassic felspars, its presence can perhaps be ignored in an elementary account of this kind. The very small amount of the Or-molecule in plagioclase can certainly be so treated, and in fact the composition of a plagioclase is almost invariably given in terms of the combined Ab and An percentages, omitting Or.

Plagioclase felspars are so important in rocks and in petrological studies that it has been found necessary to subdivide the series as follows:

Albite, $Ab_{100}-Ab_{90}An_{10}$
Oligoclase, $Ab_{90}An_{10}-Ab_{70}An_{30}$
Andesine, $Ab_{70}An_{30}-Ab_{50}An_{50}$
Labradorite, $Ab_{50}An_{50}-Ab_{30}An_{70}$
Bytownite, $Ab_{30}An_{70}-Ab_{10}An_{90}$
Anorthite, $Ab_{10}An_{90}-An_{100}$

Composition is not the sole factor which controls the nature of the felspars. Both temperature, and to a less extent, pressure have an important influence. Thus amongst potassic felspars there are three main mineral species, each stable under limited temperature and pressure conditions. **Sanidine** (Monoclinic) is formed in lavas and minor intrusions where cooling of very hot magma takes place rapidly and at a low pressure. **Orthoclase** is the stable form of $KAlSi_3O_8$ found in most igneous rocks which have cooled more slowly. **Microcline** (Triclinic) is believed to form at generally lower temperatures than orthoclase, and since it is common in many schists and gneisses as well as granites and pegmatites, it may be that its formation is aided by shearing stress. These three minerals provide a good instance of polymorphism.

Although high- and low-temperature modifications can be distinguished in the plagioclase series also, their recognition is a matter for experts.

Occurrence. Felspar is collectively the most abundant mineral in igneous rocks. It is also present in gneisses and many schists, and to a small degree in many sediments, especially felspathic sandstones and arkoses.

In acid igneous rocks such as granites, potassic felspars (orthoclase or microcline) may occur with albite or oligoclase and quartz. Sanidine occurs as phenocrysts in acid volcanic rocks. In Intermediate rocks, which are less siliceous, orthoclase is very abundant in syenites (and sanidine in trachytes), while andesine is typical of diorites and their fine-grained equivalents, the andesites. In basic igneous rocks relatively poor in silica and generally rich in CaO, MgO and FeO, calcic plagioclase is dominant and orthoclase becomes a rarity.

Potassic Felspars

It is convenient to describe sanidine, orthoclase and microcline together. The two former are almost identical in crystal form, habit and twinning. They can be distinguished by their occurrence (sanidine as phenocrysts in lavas, etc.) and associated minerals, and also by their optical orientations. Microcline is just Triclinic and can only be distinguished from orthoclase in hand-specimen by more accurate crystal measurement than a contact goniometer affords; but in thin section it is quite distinctive on account of multiple twinning, as described below.

Form and Cleavage. Orthoclase and sanidine crystals vary in shape and are generally either flattened parallel to the side-pinacoid (010), or elongated along the *a*-axis as shown in fig. 21. Cleavages are almost equally well developed parallel to the basal and side pinacoids. These cleavages are important since they provide one useful means of distinguishing orthoclase from quartz, which is otherwise rather similar in appearance in thin section.

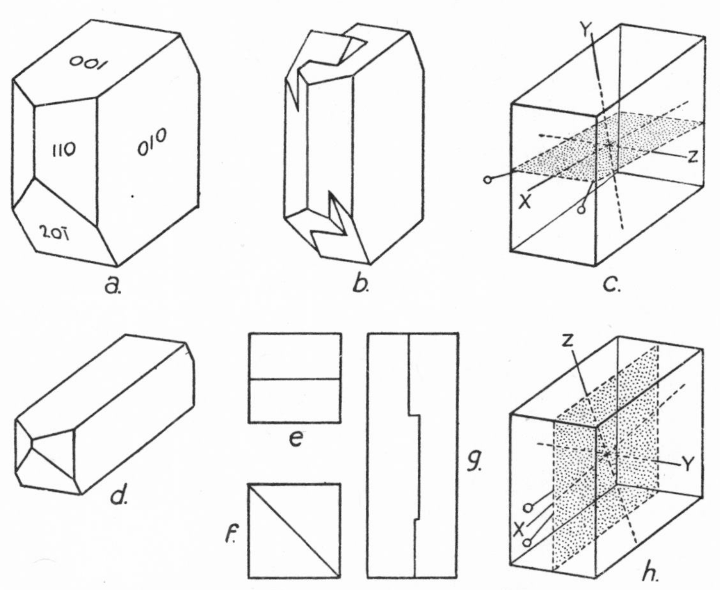

FIG. 21. *Orthoclase and sanidine (a) normal Carlsbad habit; (b) interpenetrant Carlsbad twin; (c) optical orientation of orthoclase; (d) Baveno habit; (e) section of Manebach twin; (f) section of Baveno twin; (g) section of Carlsbad twin; (h) optical orientation of sanidine. Optical axial planes stippled.*

Colour. When ideally fresh, orthoclase is a colourless transparent mineral. Generally it is altered sufficiently to make it opaque in hand-specimen, and it may be white or iron-stained to a varying degree, when it becomes buff, brown or even brick red. It is colourless in section but may be turbid through alteration.

Refractive Indices and Birefringence. For practical purposes refractive indices of sanidine, orthoclase and microcline are identical, and it is useful to know that *they are considerably lower than those of quartz* and even lower than Canada balsam. For orthoclase $n\alpha = 1{\cdot}519$; $n\beta = 1{\cdot}523$ and $n\gamma = 1{\cdot}525$. Surface relief is negligible. Birefringence (0·006) is weak giving first order greys.

Sign and Orientation. The optic axial plane in orthoclase is perpendicular to (010). 2V is large (about 70°) and orthoclase is optically negative. In high-temperature sanidine the optic axial plane becomes parallel to (010), with a small negative 2V.

Extinction. Because of the Monoclinic symmetry, sections of orthoclase cut perpendicular to (010)—including (001) and (100) sections — give straight extinction. (100) sections show both basal- and side-pinacoid cleavages intersecting at 90°. Elongation of the section is generally in the direction of the side-pinacoid cleavage. In (010) sections only the basal cleavage is visible and extinction is slightly inclined to this.

Twinning (see p. 49). This is perhaps the most distinctive feature of felspars and the one upon which rapid identification is based in thin section. Briefly, orthoclase and sanidine are only simply twinned with the halves of the twin joined along (010), (001) or a diagonal plane (0k1). The first type is most frequent and is known as twinning on the Carlsbad Law. It is interesting to note that the (010) plane along which the two parts of the twin unite—the so-called composition plane—is *not* the twin plane. Indeed it cannot be a twin plane in a Monoclinic mineral in which (010) is the normal plane of symmetry. The Carlsbad twin plane is actually (100).

Microcline shows patchily developed multiple twin lamellæ occurring in two directions which intersect at about 90°. The effect seen between crossed polarisers is highly distinctive and is best described as a 'cross-hatching'. The slender lamellæ are frequently spindle-shaped and tapering (Pl. VIII).

Alteration. Felspars readily alter to kaolin (hydrous aluminium silicate) which makes the crystals turbid with innumerable dust-like inclusions. Minute flakes of white mica (sericite) may also develop, particularly in potassic felspars. The sericite can be distinguished even in its very finely crystalline state because of its strong birefringence which gives pin points of bright colours between crossed polarisers. Although alteration may obscure the identity of a felspar, it provides a useful and rapid means of distinguishing invariably clear quartz from felspar, and often of recognising two kinds of felspar in one rock due to their different degrees of alteration.

Plagioclase Felspars (Pl. VIII)

Form and Cleavage. The external forms of plagioclase crystals are similar to orthoclase in many respects, except that the *b* and *c* crystallographic axes are no longer at right angles, but make angles of from 93° to 94°. Hand-specimen crystals are rarely available for study; but another proof of the Triclinic symmetry of plagioclase is provided by the basal- and side-pinacoid cleavages which may be seen in (100) sections to intersect at about 86°. Plagioclase felspars provide the student with his most important example of Triclinic symmetry.

Colour. As for orthoclase.

Refractive Indices and Birefringence. Refractive indices vary with composition, and their accurate measurement provides one satisfactory way of determining the albite and anorthite content of a plagioclase. The reader should consult larger textbooks for the necessary graphs and details of the techniques involved. Refractive indices range from 1·525 to 1·588 increasing in a linear manner from albite to anorthite (see fig. 23). Only albite has all its refractive indices below those of quartz. Surface relief is low. Birefringence remains fairly constant throughout the series and is slightly stronger than for orthoclase. Interference colours are first order greys and whites.

Sign and Orientation. The three principal vibration directions X, Y and Z, while remaining mutually at right angles, show a gradual shift in position relative to the crystallographic directions. This shift is dependent upon composition. The optical orientation, then, is different for each kind of plagioclase. Since plagioclase is Triclinic there are no symmetry planes to control the orientation and the problem can only be solved completely with the aid of a universal tilting stage for the microscope and stereographic projection of the determined crystallographic and optical directions—techniques which are far beyond the scope of this book.

Twinning. Plagioclase may easily be distinguished from orthoclase on account of its multiple lamellar twinning seen between crossed polarisers. A description of the predominant plagioclase twinning on the Albite Law (twin plane, 010) is given on p. 50. In addition plagioclase may have twin lamellæ which lie perpendicular to the (010) plane, according to the Pericline Law. Complex twins combining the Carlsbad and Albite Laws are quite common. The student should guard against misidentifying grains of plagioclase which show no twinning. This may be simply because they happen to be sectioned parallel to the twin planes; but in certain metamorphic rocks untwinned plagioclase (anorthite or oligoclase in different cases) is common.

Extinction and Determination of Composition. The variation in optical orientation mentioned above, combined with the multiple twinning, provides a considerable number of ways of determining plagioclase composition. One of the simplest techniques relies upon the fact that the (010) Albite twin planes become in effect planes of symmetry separating adjacent twin lamellæ. The optical directions of the latter are exactly reflected across this plane. This applies to the

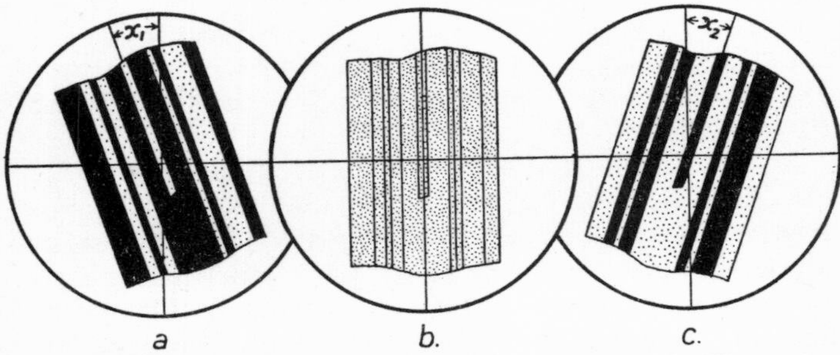

FIG. 22. *Measurement of symmetrical extinction of Albite twin lamellae in plagioclase. For explanation see text.*

extinction phenomena. Any section of plagioclase cut perpendicular to the (010) twin planes will therefore give equally inclined (symmetrical) extinction on either side of (010). The rotation necessary to put one set of lamellæ in extinction to the left of the N-S crosswire will be equalled by the rotation to the right which extinguishes light from the other set (fig. 22). If a number of measurements of this kind are made and the *maximum* angle of symmetrical extinction is found, this angle may be used to obtain an approximate composition from the curve in fig. 23.

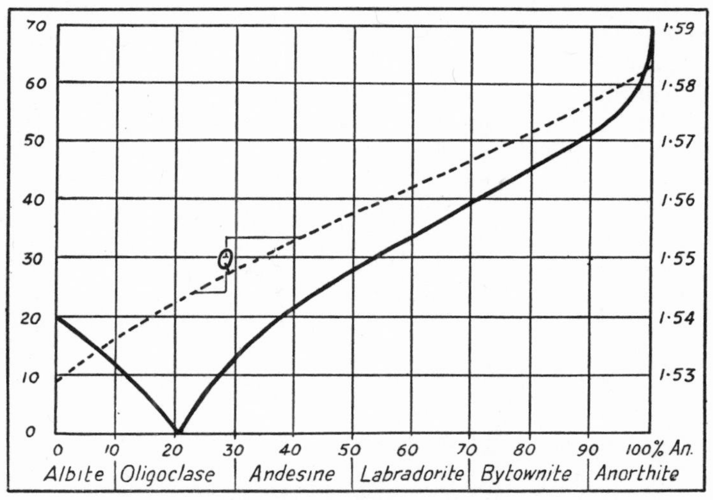

FIG. 23. *Curve of maximum symmetrical extinction angles of Albite twin lamellae of plagioclase (wavy continuous line). Variation of refractive index* β, *shown by broken line. Refractive indices of quartz, Q, are shown for comparison.*

Alteration of the more sodic plagioclases is generally similar to that of orthoclase, as described above; but the more caleic members tend to alter in a distinctive fashion involving splitting the two component parts, Ab and An. The former is relatively stable; but the latter changes into a member of the epidote group.

Zoning, which is often conspicuous in plagioclase, has been described on p. 48.

Perthites

Alkali felspars in plutonic rocks may be in the form of intergrowths of orthoclase or microcline on the one hand with a sodium-rich plagioclase on the other. In these intergrowths one mineral (generally the potassic felspar) plays the role of host to inclusions of the other. These intergrown felspars are termed perthites and are named according to the nature of the host mineral, as for example, orthoclase-perthite, microcline-perthite, or antiperthite when plagioclase is the host mineral. The inclusions are generally lamellæ, films, lenses or 'strings' evenly scattered with uniform orientation through the host. Uniform optical orientation of the inclusions may be proved by their simultaneous extinction. The origin of some of the coarser vein-like inclusions in perthites is open to question. It may be a replacement phenomenon. Some types of the finer structures, however, can be removed by heating the specimen. This produces a homogeneous sanidine-type of felspar with the albite in solid solution in the sanidine structure. Reasoning from this fact, it is presumed that slow cooling causes the albite (plagioclase) component to come out of solid solution. The phenomenon is appropriately termed ex-solution. It appears that perthites are the low temperature, slowly cooled equivalents of sodium-rich sanidines.

THE FELSPATHOIDS

As the name suggests, the members of this family are closely related to the felspars, but are *under-saturated* with silica, which means that they are incapable of existing in the presence of free silica at the time of their formation. Thus *felspathoids and quartz do not occur in the same rock section.*

The more important felspathoids are:

Leucite, $KAlSi_2O_6$, with which compare orthoclase, $KAlSi_3O_8$

Nepheline, $NaAlSiO_4$, with which compare albite, $NaAlSi_3O_8$;
Nosean, similar to nepheline, but with S;
Hauyne, similar to nepheline, with Ca and S;
Sodalite, similar to nepheline, with Cl.

Leucite. Cubic and pseudocubic (Pl. IX). A high-temperature form of leucite persists in a metastable condition to ordinary temperatures under conditions of rapid cooling and is genuinely Cubic and isotropic. Slower cooling produces a variety which crystallises as apparently perfect icositetrahedra—a well-known Cubic form; but as these crystals are seen in thin section to be complex twins and to exhibit weak birefringence, they obviously must be pseudocubic. Leucite occurs only in rocks which were quenched, the Vesuvian lavas being excellent examples. In these the leucite occurs as relatively large perfectly formed phenocrysts, and also as very much smaller second generation crystals. The former are ideally eight-sided in centrally cut sections. The refractive indices are about 1·508, and consequently the mineral has low relief. A geometrical pattern of small inclusions is characteristic. In the second generation crystals these consist of glass. As noted above, leucite exhibits very weak birefringence, the interference colour being a dull grey. Undoubtedly the most valuable diagnostic feature is the twinning, already described on p. 49 and illustrated in Pl. IX. It is important to notice that the twinning can be studied only under very good lighting conditions: it is very impressive when viewed using a good microscope lamp; but using daylight on an average winter afternoon, the twinning effects can be scarcely seen.

Nepheline, Hexagonal, polar (Pl. IX). Unlike leucite, nepheline is found not only in quickly cooled volcanic rocks, but also in deep-seated slowly cooled plutonic rocks. Euhedral crystals in volcanic rocks are easily recognised by their shapes: they occur as stumpy six-sided prisms, about as tall as they are wide. Consequently vertical sections are nearly square; the fewer basal sections are, of course, perfect small hexagons (Pl. IX). Nepheline is colourless, with low relief and weak birefringence: $n\omega$ 1·532, $n\epsilon$ 1·528. The crystals are usually cleaved, sometimes parallel to the base, but in other cases parallel to the prism faces.

In coarse-grained rocks euhedral nephelines up to perhaps half-an-inch in diameter do occasionally occur, but normally the mineral is anhedral, and, in the absence of the evidence of shape, it is much more difficult to identify: on account of lack of personality, so to speak, it may easily be overlooked or misidentified. Points of significance include straight extinction in vertical sections; but in the absence of clearly defined cleavages, it may be very difficult to see any trace of internal structure to which the extinction may be referred. The best plan is to turn the stage until the suspected nepheline is exactly in extinction, and then look carefully for the faintest traces of structure parallel to the crosswires. Further, once the crystallographic orientation has been inferred — presumably from considerations of shape or cleavage, it is easy to check the uniaxial, optically negative character of the mineral. Lastly, the associates of nepheline are helpful in this connection. As might be anticipated from its composition, nepheline occurs in soda-rich igneous rocks, and is found in association with alkali-felspars typically, with aegirine or aegirine-augite among the pyroxenes, and $Na^{+}Fe^{+++}$ amphiboles such as riebeckite. Nepheline-bearing rocks are often associated in the field with carbonatites; and therefore calcite is a not uncommon associate of nepheline. If the two occur in juxtaposition a reaction rim of cancrinite may occur between the two. This is a yellow mineral in the mass, but colourless in thin slice, with strong birefringence and bright interference colours. When present, it is the cancrinite which 'catches the eye', and which directs attention to the probability of nepheline being present.

The other felspathoids are less frequently encountered than the two described above.

Nosean occurs only as small crystals, a fraction of an inch in diameter, in certain lavas, but is nevertheless a very distinctive mineral in thin section (Pl. IX). Like leucite, all the crystals have the same shape — that of the Cubic rhombdodecahedron. Ideally therefore central sections should be six-sided; but in the majority of cases corrosion, often intense, has gone far to destroy the characteristic form. Presumably incipient magmatic corrosion is responsible for a second distinctive feature of nosean—a thick dark border of inky black or brown material. Internally crystals are pale buff in colour, and are perfectly isotropic. Cleavage traces, often emphasised by incipient alteration, are usually observable.

Hauyne closely resembles nosean: it has the same crystal form, shows the same magmatic corrosion, is isotropic; but at its best is attractively coloured light sky blue.

Sodalite is a difficult mineral to identify by optical means alone. It is colourless with a relatively strong negative relief as the refractive index is only 1·487. This feature and its isotropism provide the surest indication of its presence. The sodalite may be most clearly seen if the light is rather heavily diaphragmed down, and the microscope tube lowered a little. In these conditions the light seems to be concentrated in the areas of sodalite, while adjoining felspars appear dull by comparison. The sections are often much cleaved. The mineral analcite is all but identical with sodalite in thin section.

THE ZEOLITES

Under this name are grouped several minerals which share the common feature of easily parting with water of crystallisation on heating—hence the name, from a Greek root meaning 'I boil'. It is quite impossible in a small space to give a summary of the individual zeolites, and two only have been selected for this account.

Natrolite is one of a number of fibrous zeolites, the composition of which may be represented by $Na_2Al_2Si_3O_{10}.2H_2O$. It is therefore close to nepheline on the one hand, and to albite on the other. Terminated crystals—which are all too rarely available—*appear* to belong to the Tetragonal system and to consist of a simple combination of an elongated prism capped by a bipyramid in zone with it. Careful measurement has shown, however, that the crystals are actually Orthorhombic, though within an ace of being Tetragonal.

In thin section natrolite is colourless, it exhibits negative relief as the indices are about 1·48. Birefringence is slightly stronger than that of quartz—the optical sign (positive) is the same. It will be realised, perhaps, that some danger may arise of confusing natrolite with fibrous chalcedony; but with careful attention to the optical details this should be avoided. Both minerals may be found in vesicles in lavas.

Analcite, $NaAlSi_2O_6H_2O$ Cubic. This is the only zeolite which ranks as a rock-forming mineral having the status of an essential constituent of certain Intermediate and Basic igneous rocks. Like all the other zeolites it also commonly occurs as a secondary mineral formed from felspars or felspathoids of appropriate composition.

In shape analcite is identical with leucite; but if the important differences in occurrence are appreciated, this should occasion no difficulty. Leucite invariably occurs as scattered phenocrysts of a high temperature origin. Analcite commonly lines cavities, but leucite never occurs in this way. In thin section analcite is difficult to identify by optical means: it is colourless, normally isotropic, often with cleavage traces, and with negative relief (refractive index 1.48). It will be realised that there is little of diagnostic value in these characters. Normally analcite is formed late in the crystallisation sequence, and therefore occurs as interstitial patches between the earlier-formed crystals of plagioclase, for example, with which it is commonly associated. Actually the association may take another form, for the analcite may actually *replace* the substance of the plagioclase, penetrating initially along the cleavages. Such analcitised plagioclases appear much broken up in plane polarised light, but with crossed polarisers the veins and patches of isotropic material may readily be seen and may with confidence be identified as analcite. One helpful point may be noted. If such an interstitial patch of analcite impinges upon an augite crystal, the latter may be rimmed with bright green aegirine or aerigine-augite, formed by reaction between the original pyroxene and the strongly sodic solutions from which the analcite subsequently crystallised.

CHAPTER 10

Metamorphic and Accessory Minerals

In this final chapter are grouped those minerals which one may regard as characteristic of, though not necessarily restricted to, metamorphic rocks: the garnets, aluminium silicates, corundum, cordierite and spinels. With these are considered the products of pneumatolysis, formed through the activities of a gas-phase; and the more normal accessories.

GARNETS

Composition. The general formula may be expressed as $R^{++}{}_3R^{+++}{}_2SiO_{12}$, in which the divalent R^{++} may be Mg, Fe^{++} or Mn; while the trivalent R^{+++} is Al, Fe^{+++} or Cr. The commoner members of the group are:

Almandine (common garnet), ideally $Fe_3Al_2Si_3O_{12}$;
Pyrope (precious garnet or Cape ruby), $Mg_3Al_2Si_3O_{12}$;
Grossularite (name based on the Latin for gooseberry), $Ca_3Al_2Si_3O_{12}$;
Andradite or melanite (black garnet), $Ca_3Fe_2Si_3O_{12}$.

In this isomorphous group atomic substitution results in garnets of mixed composition, which may nevertheless be expressed in terms of the ideal 'molecules' given above. In particular Mg, Mn and Fe^{++} are mutually replaceable, as are Al, Cr and Fe^{+++}. Thus the actual composition of a natural garnet must be determined by analysis: the optical properties seen in a thin section give little indication of the composition, although it may be inferred, broadly, from the mineral assemblage.

Occurrence. Garnets are largely metamorphic in origin and are most commonly encountered in garnet-mica-schists, in which they occur as large, well-formed crystals, termed 'porphyroblasts' as they have grown in the solid. As a direct consequence of their mode of origin such garnets— normally **almandines** — are often crowded with small inclusions representing material that they were unable to assimilate. A long period of growth is sometimes proved by the occurrence of spiral zones of these inclusions which justify the name 'snowball garnets'.

Pyrope occurs as a somewhat rare accessory in very basic igneous rocks, and is characteristic of, and obvious in, the rock type eclogite. **Melanite** also occurs in igneous rocks in which it is often associated with felspathoids. **Grossularite,** on the other hand, is exclusively metamorphic and occurs in calc-schists and marbles that were originally impure limestones. The mineral association is therefore completely different: the grossularites are associated with calcite and such lime-silicates as zoisite, idocrase (vesuvianite), wollastonite and diopside.

Forms. Garnets belong to the highest class of symmetry in the Cubic system. The well-known and much-used garnet crystals in teaching collections are almandines and melanites occurring as rhomdodecahedra and icositetrahedra, alone or in combination. Central sections will appear six- or eight-sided respectively (Pl. X). Cleavage is not normally developed, but an irregular fracture is characteristic. Detrital garnet grains, on the other hand, as a result of severe buffetting, may show a complicated dodecahedral cleavage which controls the shapes of such grains.

Refractive index is high, and the relief is consequently pronounced. As the garnets are often relatively large, they can be seen easily in a slide before it is placed on the stage, by merely looking through it towards a window. In this way a faint pink colour may often be seen, though this may almost or quite disappear under the microscope.

The distinctive properties of specific garnets:

Almandine, purplish red to reddish black in the hand-specimen, is very light pink in thin slice. Refractive index is high, even for garnets—1·83.

Pyrope, rich red in hand-specimen, is also very pale pink, or even colourless in thin section, and has a somewhat lower R.I.—1·705. It is completely isotropic.

Grossularite in hand specimens is typically green, though the variety cinnamonstone is reddish brown. In thin section the mineral is invariably colourless, with R.I. of 1·735. Much of the grossularite occurring in lime-silicate rocks shows weak birefringence and fails to extinguish completely as a genuinely isotropic Cubic mineral would be expected to do.

Andradite (melanite) is black in hand-specimens (hence the second name), but in thin section is brown. Not infrequently growth zoning is pronounced, while optical anomalies—twinning and feeble birefringence lead one to suspect its Cubic symmetry.

CORUNDUM

Composition. Oxide of aluminium, Al_2O_3. Trigonal

Occurrence. As might be expected, corundum occurs chiefly in metamorphic rocks formed from argillaceous sediments, particularly when the agent of metamorphism has been very high temperature. It therefore occurs typically in argillaceous inclusions caught up in basic (basaltic) magma, and raised to near-magmatic temperatures. Its associates include cordierite, spinel, sillimanite and anorthite. In addition corundum is a rare component in certain igneous rocks, which of necessity must be deficient in silica. In this association corundum occurs with nepheline in coarse-grained syenitic rocks.

Form and Physical Features. Euhedral corundum crystals appear at first sight to be Hexagonal; but closer inspection will usually result in the discovery of forms such as the rhombohedron, and surface markings which prove the mineral to belong to the Trigonal system. The crystal habit varies considerably. Commonly corundum occurs as six-sided prisms or rather acute bipyramids capped by basal pinacoids, giving them a crude barrel-like appearance, but a tabular habit is sometimes developed. This occasions some little difficulty in thin sections, as vertical sections of such crystals are *elongated along a lateral axis*. If this were mistaken for the *c*-axis, and tested with the accessory plates, the wrong optical sign would be obtained.

Refringence and Birefringence. The refractive indices are high, about 1·76 to 1·77, and the relief is therefore strong. The birefringence, on the other hand, is weak, about 0·0082, and therefore the range of interference colours is practically the same as that of quartz. As regards colour, corundum of gemstone quality is well known as ruby and sapphire, the colours of which need no description. In thin slice the colour is usually very pale — the only British occurrences are distinctly blue — and appreciably pleochroic. Basal sections, of course, yield a uniaxial interference figure of negative sign.

ALUMINIUM SILICATES: ANDALUSITE, SILLIMANITE AND KYANITE

These three minerals of precisely the same composition but with different physical attributes, provide an example of polymorphism: the formula for all three is Al_2SiO_5. They are practically confined to metamorphic rocks of pelitic composition, though the style of metamorphism is different in each case. Andalusite occurs in thermally altered argillaceous rocks, where the temperature achieved was not high; but where the temperature was higher, sillimanite takes its place. By contrast kyanite is normally a product of high grade dynamothermal (regional) metamorphism, and often occurs closely associated, and even intergrown with, staurolite.

Andalusite. Orthorhombic, forming euhedral prismatic crystals with a rhomb-shaped, nearly square cross-section (Pl. XI). Sometimes grain-boundaries are less well-defined, and the andalusites may be charged with inclusions of biotite, iron-ore, quartz, etc. Prismatic cleavage may be conspicuous. In the variety **chiastolite** (Pl. XI), black, probably carbonaceous inclusions are concentrated in a distinctive manner, giving the effect of a whitish Maltese cross against a black background, when the crystal is seen in cross-section. These chiastolites vary in different specimens from hair-like needles to sturdy crystals an inch or so in diameter, embedded in a little-altered slaty matrix.

Diagnostic features include moderate relief, combined with weak birefringence: $n\alpha$ 1·635, $n\gamma$ 1·645; prismatic cleavage, and under favourable conditions a distinctive, though weak, pleochroism, from very pale pink to colourless. Detrital grains, which are considerably thicker than a normal good section, show a much more pronounced change and the rich pink colour is distinctive, and almost diagnostic in itself. The mineral is optically negative, biaxial, with large 2V.

Sillimanite. Orthorhombic, forming long prismatic, often needle-like crystals, which may be aggregated into fibrous masses, sometimes felt-like ('fibrolite'). Rather nondescript fibrous aggregates, closely associated with mica in schist is characteristic; though larger single crystals also occur embedded in quartz or cordierite in some types of sillimanite-gneiss.

Sillimanite crystals are colourless, with moderate relief and birefringence: $n\alpha$ 1·66, $n\gamma$ 1·68, theoretically giving interference colours low in the second order; but actually these colours are only seen when several fibres are in parallel orientation. Extinction is, of course, straight.

Kyanite (Pl. XI). Triclinic, usually elongated parallel to the *c*-axis, and flattened parallel to (100) so that crystals appear to be thin (nearly) rectangular tablets. An irregularly distributed attractive sky-blue colour is highly characteristic. Kyanite has three cleavage directions: the most perfect being parallel to the front-pinacoid, while that parallel to the side-pinacoid is only slightly inferior. The third, parallel to the basal-pinacoid, is more widely spaced, a parting rather than a true cleavage, and is non-rectangular relative to the other two. Detrital kyanite is one of the easiest 'heavy minerals' to identify as the grain shape, controlled by the cleavage, is distinctive. In thin section, however, diagnosis of a particular grain may be less positive. The sections are colourless, much cleaved, with moderate to strong relief, and show oblique extinction to the cleavage traces (Pl. XI). The maximum angle of about 30° occurs in (100) sections which yield a fairly well-centred Bx_a interference figure with large 2V. Refractive indices, nα 1·717, nγ 1·729.

STAUROLITE

An Orthorhombic, holosymmetric mineral, perhaps best known in the form of brownish cruciform twins. Staurolite is less widely distributed than andalusite and kyanite, though like these minerals, it occurs in pelitic schists and gneisses. Indeed kyanite and staurolite occur frequently in the closest association, and may be intergrown, with a perfect staurolite prism occurring centrally within a bladed kyanite. In composition staurolite is essentially an iron-bearing kyanite.

In thin section staurolite is yellow to yellowish brown, and distinctly pleochroic, with strong relief. Refractive indices about 1·74 to 1·75. Prismatic cleavage occurs, and is important in detrital staurolite as it gives a characteristic saw-like edge to the mineral grains.

CORDIERITE

This mineral is essentially an aluminous silicate of Mg and Fe and may conveniently be thought of as being compounded of two spinel and five silica 'molecules'. Thus $2(Mg,Fe)O.Al_2O_3$ plus $5SiO_2$ gives $(Mg,Fe)_2Al_4Si_5O_{18}$—an approximate formula—and thinking in these terms serves to underline a fact of some importance in the identification of this difficult mineral — the almost invariably close association of cordierite with spinel.

Occurrence. Cordierite is wide-spread as an important constituent of thermally altered argillaceous rocks of hornfels type. It also occurs in certain contaminated igneous rocks of both acid (granitic) and basic composition, notably in contaminated norites (basic rocks consisting essentially of plagioclase and hypersthene).

Form and Other Physical Features. Orthorhombic, nearly Hexagonal. Well-formed crystals are seldom encountered, though euhedral pseudomorphs after cordierite, obtained from some of the West of England granites are seen in mineral collections. Massive cordierite has the transparency and lustre of quartz and the deep violet colour of amethyst, but is pleochroic. In thin section cordierite is distressingly like quartz in several important respects, and it is only by paying meticulous attention to details that the mineral can be positively identified.

Refractive indices are $n\alpha$ 1·540, $n\gamma$ 1·549 (compare quartz with 1·544 and 1·553 respectively). Thus the relief is not appreciably different, while the birefringences of the two are identical (0·009), and so therefore is the range of interference colours. As regards points of difference: cordierite *may* show a feeble cleavage, and is chemically less stable than quartz. It may show alteration, starting along, and tending to emphasise, the cleavage, into an aggregate of chlorite and white micaceous mineral, termed 'pinite'. Apart from these differences, even if the cordierite is quite fresh and uncleaved, there are three points of significance. Firstly, the close association of the mineral with spinel, which occurs usually as aggregates of small very dark green to opaque octahedra embedded in the cordierite. Secondly, light yellow haloes surrounding minute inclusions are characteristic, though they can be seen only under good lighting conditions. Thirdly, a suitably orientated section may show distinctive sectorial twinning; though it must be carefully noted that such a twin, in *vertical* section, would show a number of twin-bands parallel to the crystal edges, which gives them a close resemblance to twinned felspars.

Summarily, cordierite may be highly distinctive in thin section and easily identified on sight; but on the other hand it may tax the expert. It may happen, of course, that a suitably orientated grain may yield unequivocal evidence in the shape of a biaxial interference figure, optically negative, with large, variable 2V.

THE ACCESSORY MINERALS

An ordinary igneous rock is named and classified largely on the basis of the kinds and proportions of the essential minerals present. Take out any one of the essential minerals and the name and classification are altered. But in addition to these essential minerals, most rocks contain small quantities of other, less important, components the presence or absence of which is deemed to make no difference to name or classification. These are the accessories, certain of which are rarely absent from igneous rocks of most kinds. Among the most widely occurring are apatite, sphene, zircon and one or other of the iron-ores.

Apatite. $Ca_5(F,Cl)P_3O_{12}$. Hexagonal, usually a simple combination of prism and basal pinacoid (Pl. II).

Occurrence. It appears probable that apatite occurs in all igneous and most metamorphic rocks, seldom in large amounts, but it is very seldom that a section fails to show some apatite. Although apatite crystals of large size occur in some pegmatites, particularly those of syenitic composition, the petrologist is far more familiar with the microscopic Hexagonal prisms scattered through the slide and which are easily identified as apatites (Pl. II). The mineral has an inherent tendency towards euhedrism; and the ratio of length to girth is such that it is appropriate to speak of some of the crystals as 'needle apatites'.

Diagnostic Features. Apart from shape, which goes far towards the identification of apatite, the absence of colour, moderately high relief (n_ω 1·634, n_ε 1·629) and weak birefringence are helpful. A grey interference colour is invariable, even for most detrital apatite; vertical sections are length-fast (optically negative) and naturally show straight extinction. The six-sided basal sections are, of course, isotropic. Cored apatites are not uncommon, while careful search will often disclose hollow crystals of this mineral.

Sphene, *sometimes called titanite.* $CaTiSiO_5$. Monoclinic (Pl. II). Sphene is a wide-spread accessory, particularly in Intermediate igneous rocks such as syenites and diorites, but is equally common in certain metamorphic rocks, particularly in hornblende-schists and amphibolites. In both groups of rocks the association of common hornblende with sphene is noteworthy. Well-formed crystals are wedge-shaped in thin section and resemble a somewhat drawn-out ace of diamonds (Pl. II). In metamorphic rocks the habit is often different: the sphene occurs in the form of small lozenges in hornblende. The distribution of the sphene often shows it to have originated by the breakdown of a titanium-rich silicate, as a consequence of falling temperature or of the onset of metamorphic conditions.

Diagnostic Features. The characteristic shape, clove brown, reddish brown or greyish brown colour and slight pleochroism, well-developed cleavage traces, usually not parallel to the bounding edges, and simple twinning are common features. In addition the relief is exceptionally high ($n\alpha$ 1·90, $n\gamma$ 2·09), while the birefringence is stronger than that of ordinary rock-forming minerals with the exception of the natural carbonates. Consequently the interference colours are exceptionally weak—the so-called 'high order whites'. This fact provides a unique test for sphene: if the grain is placed in the 45° position (half-way between the extinction positions), no difference will be observed on crossing and uncrossing the polarisers, for a self-evident reason. The addition of white (interference colour) to brown (body colour) gives brown, of course.

Zircon. $ZrSiO_4$. Tetragonal, holosymmetric. Although zircon is seldom encountered in a chance section, the mineral is widely distributed in small amounts in many igneous, metamorphic and sedimentary rocks. It is chemically stable and physically durable, so that it is passed on from an original igneous source-rock through successive sediments, suffering no change except in shape as it becomes progressively more rounded. Zircon is probably best known as a ubiquitous component of 'heavy mineral suites' separated from arenaceous sediments. In these it can be studied to greater advantage than in most igneous rocks.

Diagnostic Characters. Zircon occurs as perfect, simple Tetragonal prisms capped by bipyramid faces. Although buff to brown in large crystals, under the microscope they are normally quite transparent, and colourless. The refractive indices are very high—1·92 or more—so that the relief is exceptionally high. The birefringence is also strong, and extinction straight. As a consequence of the thickness of detrital zircons, very high order interference colours are displayed centrally, but the lower order colours form colour-fringes round the crystal margin. If a fairly high magnification is used, by counting the number of such fringes, the order of the interference colour at the centre may be determined.

The distinctive pleochroic haloes which surround zircons embedded in biotite have been already described (p. 36).

Rutile. TiO_2. Tetragonal, holosymmetric. Rutile is an occasional accessory mineral in some igneous and metamorphic rocks. Like zircon, on account of its chemical and physical stability it often figures prominently in heavy-mineral suites separated from sediments. In mineral collections specimens of quartz containing numbers of rutile prisms are not uncommon. The latter may be as thick as a pencil, but are much more commonly slender needles of a typical reddish brown colour. In thin sections of many granites the quartz grains are not infrequently seen to be crowded with exceedingly minute hair-like inclusions, which, by analogy with megascopic specimens mentioned above, are inferred to be rutile, though they cannot be identified by the usual optical tests. Similarly some specimens of mica, particularly those which display the phenomenon of asterism, are seen under the microscope to contain regularly disposed hair-like crystals, arranged in planes on a pattern consistent with the pseudo-Hexagonal symmetry of the mica. These also are inferred to be rutile, doubtless exsolved from the mica.

When the crystals or grains are large enough to be studied optically, it is clear that the mineral is distinctive by reason of its colour in thin section—almost a terra-cotta red, and 'fox-red' in detrital grains. By *reflected* light a bright, almost metallic sheen is observed. Rutiles have excessively heavy outlines in consequence of the high refractive index—about 2·8. The birefringence, 0·3, is the highest of any rock-forming mineral.

Cassiterite. SnO_2. Tetragonal, holosymmetric, the crystal forms and types of twinning being identical with rutile. In hand specimens the two minerals are easily distinguishable by differences in colour and lustre; but small detrital grains may be so alike that a microchemical test has to be applied for positive identification. Cassiterite is a relatively rare accessory, restricted to granitic rocks, especially pegmatitic and pneumatolytic facies. The only justification for including cassiterite for consideration in a book of this scope is its well-known occurrence as a tin ore in Cornwall.

At their best cassiterite crystals are squat, dark brown, highly lustrous prisms, capped by bipyramids, and not infrequently twinned. In thin section and detrital grains, a distinct colour zoning, illustrated in Pl. XII, is often seen, the colours involved being shades of brown and red to colourless. As noted under rutile, a microchemical test is sometimes necessary to distinguish between that mineral and cassiterite; but normally the nature of the mineral association provides evidence. Rutile is associated with staurolite, kyanite and garnet; but cassiterite occurs with pneumatolytic and pegmatitic minerals such as tourmaline, topaz and fluorite.

Topaz. Composition similar to andalusite, but with one oxygen atom replaced by fluorine and the latter in part by hydroxyl: $Al_2(F,OH)SiO_4$. Orthorhombic, holosymmetric. Topaz is a rare mineral of restricted occurrence, found only in granitic rocks, especially those of pegmatitic and pneumatolytic types. It may form well-shaped crystals, much prized by the collector, and varying in colour from completely colourless to honey-brown. Topaz of gemstone quality and of the latter colour occurs in Brazil and is well known. Very faintly bluish to colourless topaz is much more widely distributed. Under the microscope the mineral lacks distinctiveness. It is completely colourless, and polarises with the same range of interference colours as quartz, felspars (and basal sections of white mica), all of which are its associates in rocks. Its relief is noticeably higher than quartz and felspar as the refractive indices are about 1·62. A valuable feature to be looked for is a perfect basal cleavage. As the optic axial plane is perpendicular to (001), cleavage flakes give a Bx_a interference figure, with a 2V of about 65°.

Fluorite. CaF_2. Cubic, holosymmetric. As noted above, this is one of the pneumatolytic group of minerals, often seen in close association with tourmaline, topaz and cassiterite. It also occurs in parts of Britain as a gangue mineral in mineral veins, notably in the zinc-lead mining area of the northern Pennines.

The optical characters are simple in the extreme. The fluorite in rock sections is usually anhedral, so the evidence of the well-known Cubic forms shown by the beautifully crystalline specimens obtained from vugs (cavities) is not available. The mineral has, however, a perfect octahedral cleavage, which is usually conspicuous, and a marked surface relief. It is important to note that this is a *negative relief,* as the refractive index, 1·43, is considerably below that of the embedding media in common use. The mineral is, of course, completely isotropic. It might be expected from the appearance of mineral specimens of fluorite that it would be colourless in thin slice. Actually this is often the case, but some grains show a patchy distribution of quite strong violet colour, the general appearance being reminiscent of ink-smudging.

Tourmaline, a very complex borosilicate in different varieties of which Na, Ca, Li, Al, Fe, Mg and other elements occur in varying proportions.

Trigonal. The crystallographic features are of particular interest as tourmaline belongs to a polar class of symmetry. This fact is often strikingly displayed by the development of dissimilar faces at the two ends of the crystal, and by a curious unsymmetrical colour-variation along the length of the crystal: thus certain tourmalines of gemstone quality are clear bluish green for about half their length, the other half being bright pink. The dominant form is a Trigonal prism, often modified by a Hexagonal prism at the edges. Basal sections are typically triangular, therefore, though they may be six-sided.

Occurrence. Tourmaline occurs in greatest abundance in 'acid' igneous rocks, particularly in granites which experienced a well-marked pegmatitic and pneumatolytic phase of crystallisation. In these circumstances it is commonly associated with muscovite, topaz, fluorite and, more rarely, cassiterite. Tourmaline is also widely distributed in schists, and is usually present in heavy-mineral suites separated from sediments.

Diagnostic Characters. Tourmaline displays a well-marked tendency towards euhedrism: the crystals are well formed and vary from prismatic to acicular in habit. Quite commonly the crystals form parallel or radial aggregates, which in extreme cases become spherulitic. There is no cleavage, but a rough basal fracture is commonly developed. The colour phenomena are striking, and have already been referred to (p. 34). Tourmaline containing iron is dark green to black in hand specimens, and often quite strongly coloured in thin section. Pleochroism is strong, and is undoubtedly the most striking feature of tourmaline under the microscope. The ordinary ray (ω) is much more strongly absorbed than the extraordinary ray (ϵ). Indeed in detrital grains and thick slices the absorption may be complete, so that in one position of the stage the grain is quite black. This difference in absorption may be used to verify the vibration direction of the lower polariser in the microscope, since a tourmaline prism is darkest when it lies with its *c*-axis at right angles to the vibration direction of the polariser. Basal sections always show the darkest colour, and are, of course, non-pleochroic. The colours displayed vary widely: they may be blue, green, rather muddy brown ranging down to light buff and even colourless in the position of minimum absorption. Zoning is characteristic: the colour differences displayed in some cross sections are spectacular.

The features described above are so distinctive that it is seldom necessary to check the identification of tourmaline by the standard optical tests; but when the need arises it is easy to show that the prismatic sections are length-fast, while basal sections frequently show a uniaxial interference figure of negative sign. Refractive indices vary with composition, but all kinds of tourmaline show moderate to high relief, and moderate birefringence.

THE EPIDOTE GROUP

Only two of the several minerals belonging to this group are sufficiently common to justify inclusion in this book. They are zoisite, which is the only Orthorhombic member of the group, and epidote itself, which is Monoclinic. Both are widely distributed in metamorphic rocks and in altered igneous rocks, especially those of basic composition.

Zoisite, $(OH)Ca_2Al_3Si_3O_{12}$, has a prismatic habit. Cleavage is perfect, parallel to (010). Thin sections are invariably colourless, with moderately high relief (refractive indices about 1·70). The most striking feature by which the mineral is often recognised is weak birefringence, with such strong dispersion that suitably orientated sections fail to

PLATE I

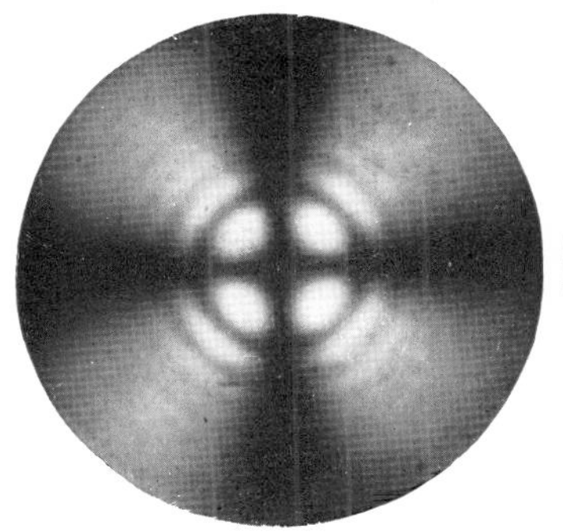

1. Uniaxial interference figure.

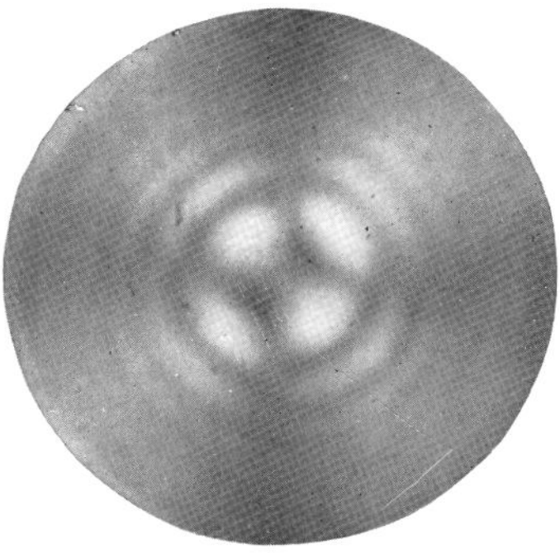

2. Uniaxial interference figure. Mica plate introduced.

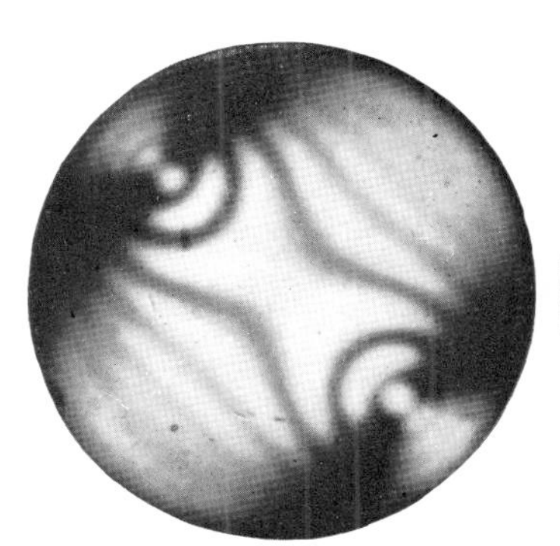

3. Biaxial interference figure. Oblique position.

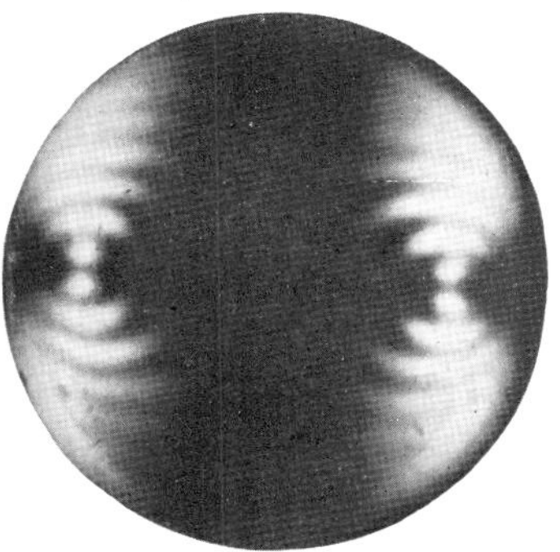

4. Biaxial interference figure. Straight position.

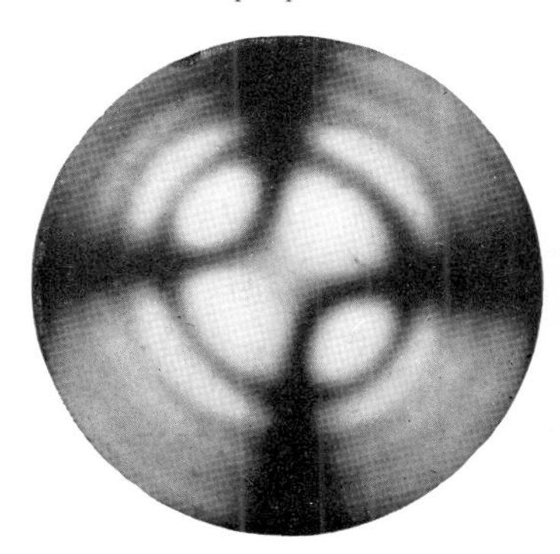

5. Pseudo-uniaxial interference figure. Oblique position.

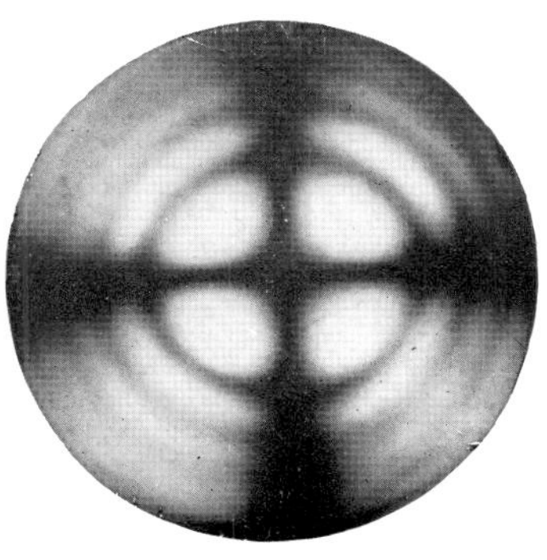

6. Pseudo-uniaxial interference figure. Straight position.

PLATE II

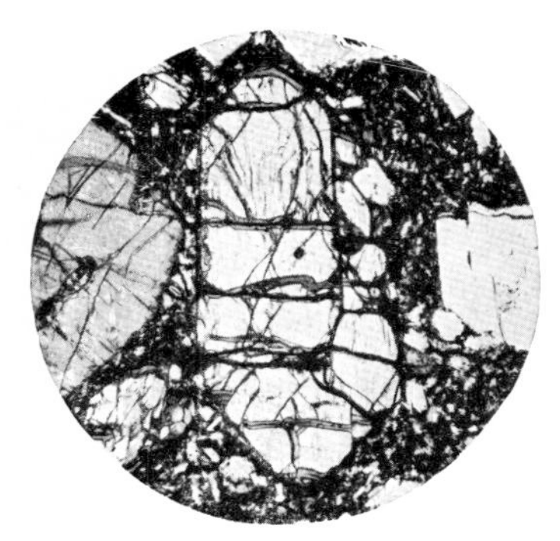

1. Olivine. Showing form and alteration.

2. Olivine. Showing form and more advanced alteration.

3. Serpentine. Pseudomorph after Olivine.

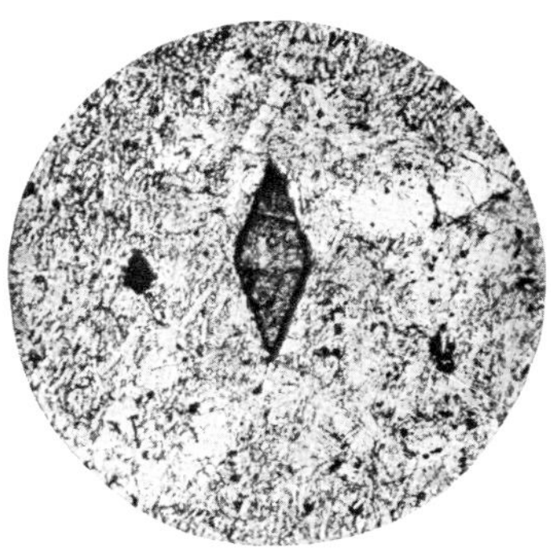

4. Sphene. Showing form and high refractive index.

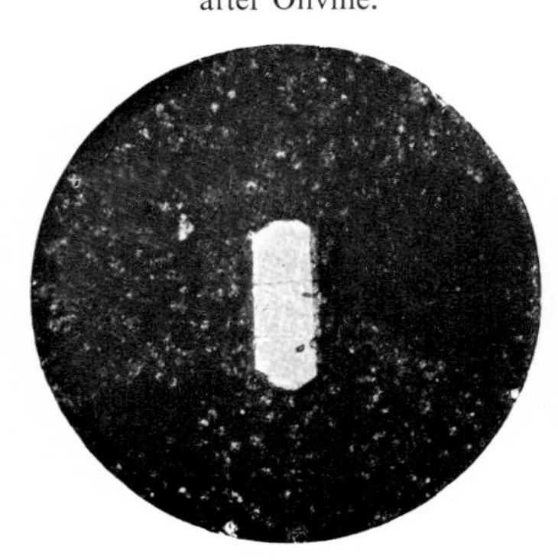

5. Apatite, vertical section. Showing form and cleavage.

6. Apatite, basal section. Showing form.

PLATE III

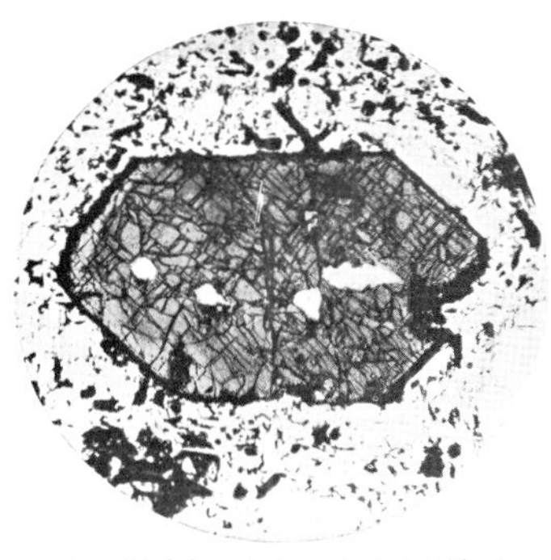

1. Ægirine, transverse section. Showing form and cleavages.

2. Augite, vertical section. Showing form.

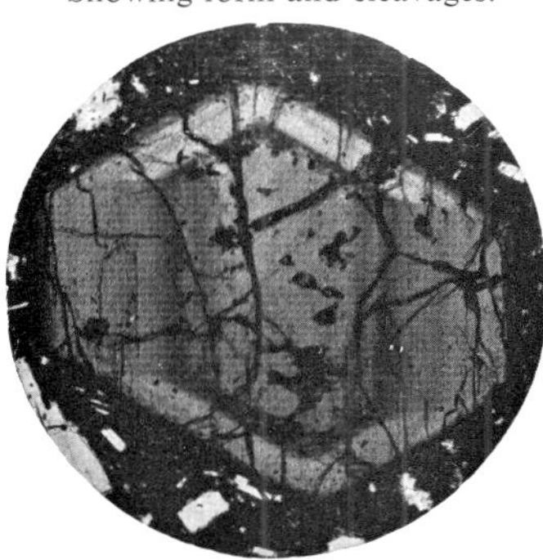

3. Augite, vertical section. Showing form and zoning.

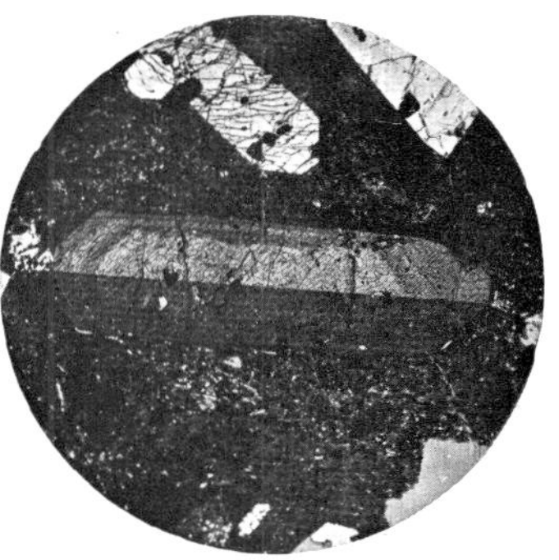

4. Augite, trans. sect. Crossed polarisers. Showing form zoning and simple twinning.

5. Augite. Crossed polarisers. Showing zoning and 'hour-glass' structure.

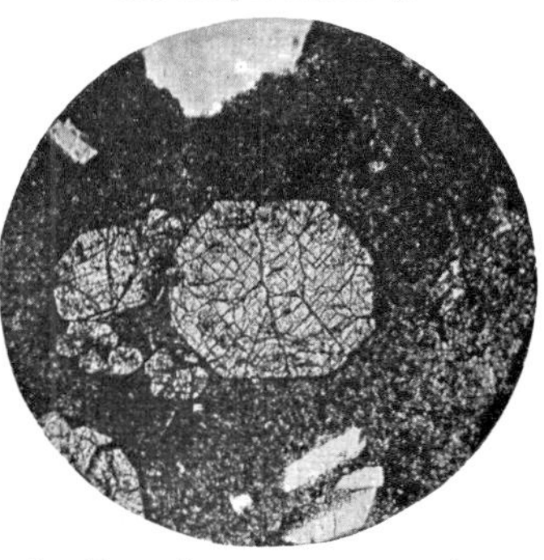

6. Enstatite, transverse section. Showing form and cleavages.

PLATE IV

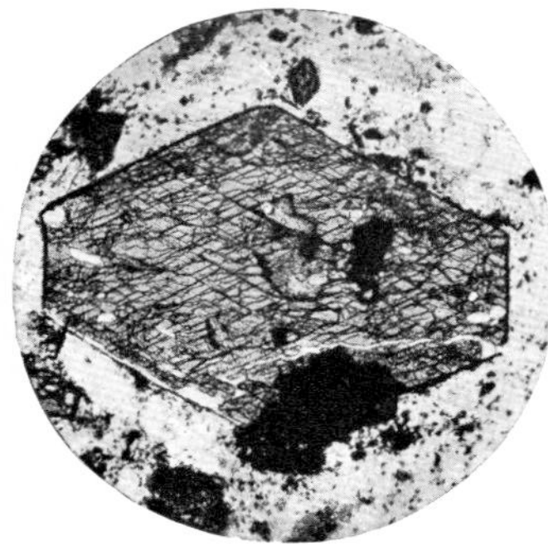

1. Hornblende, transverse section. Showing form and cleavages.

2. Actinolite, transverse section. Showing form and cleavages.

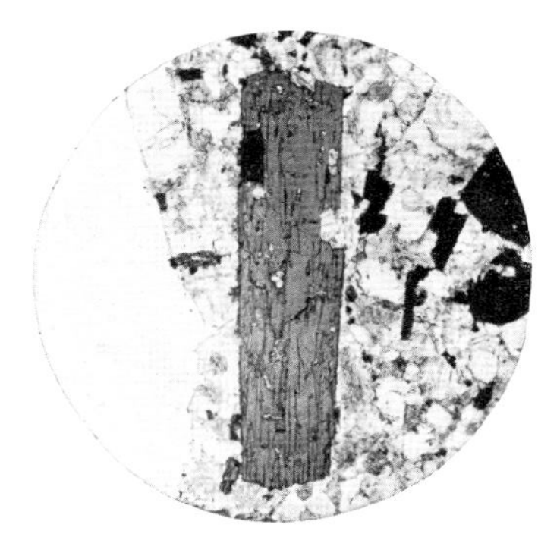

3. Hornblende, vertical section. Showing form and cleavage.

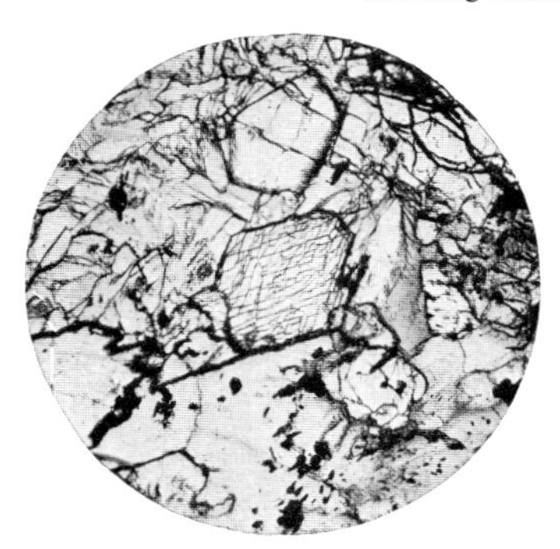

4. Glaucophane, trans. sect. Showing form and cleavages.

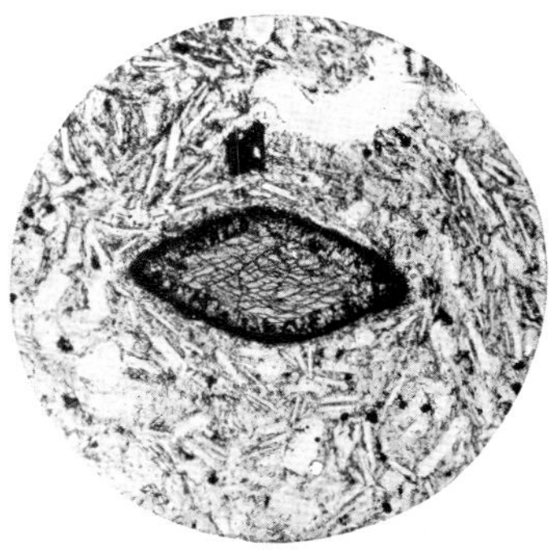

5. Hornblende, trans. sect. Showing alteration rim.

PLATE V

1. Muscovite. Showing cleavage.

2. Muscovite. Showing cleavage intergrowth with biotite.

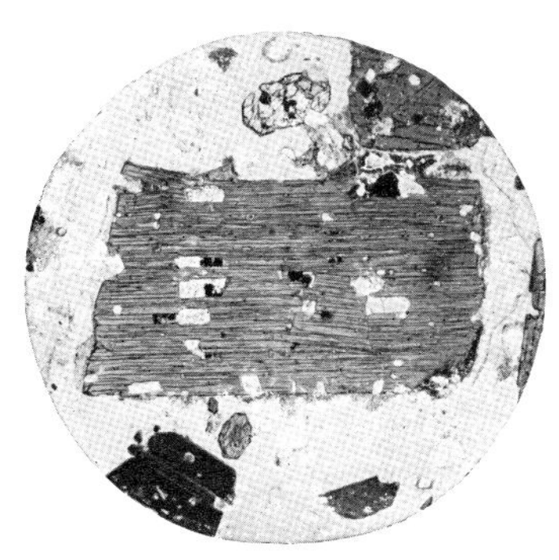

3. Biotite. Showing form, cleavage and inclusions.

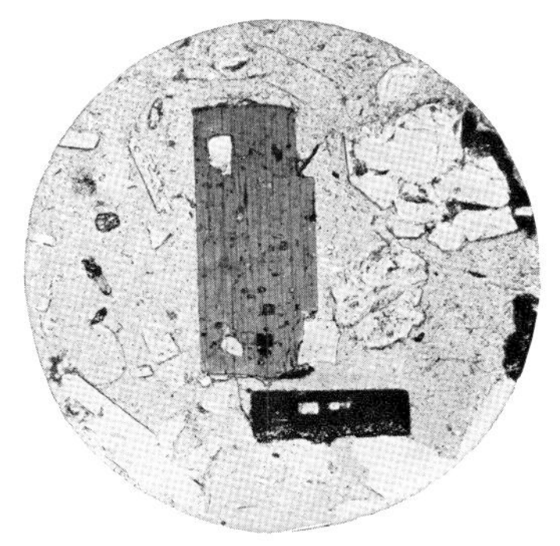

4. Biotite. Showing pleochroism.

5. Biotite. Showing partial alteration to Chlorite.

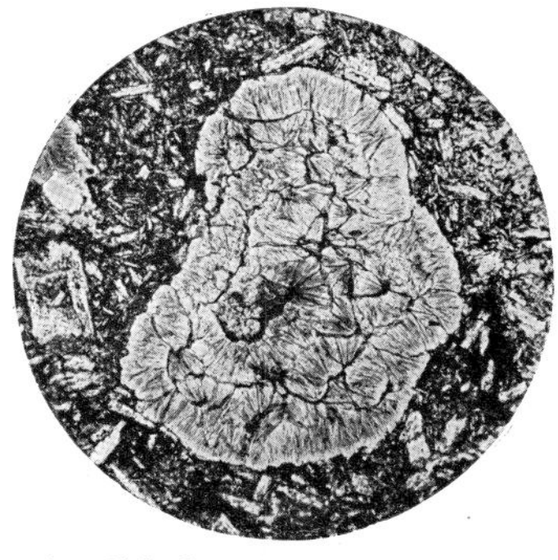

6. Chlorite. Showing vermicular habit and occurrence in amygdule.

PLATE VI

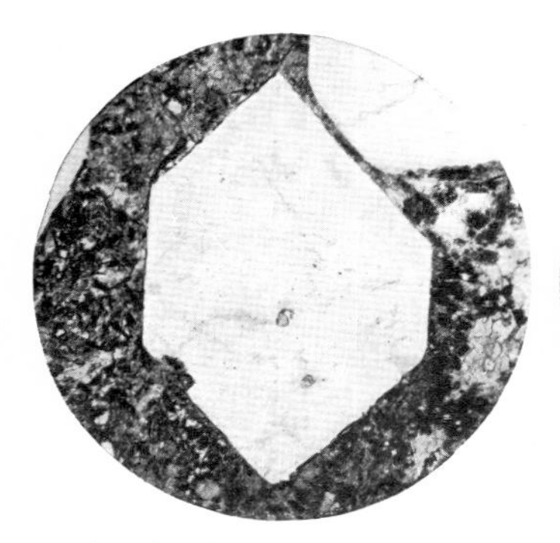

1. Quartz. Vertical section. Showing form.

2. Quartz. Basal section. Showing form.

3. Quartz. Showing corrosion.

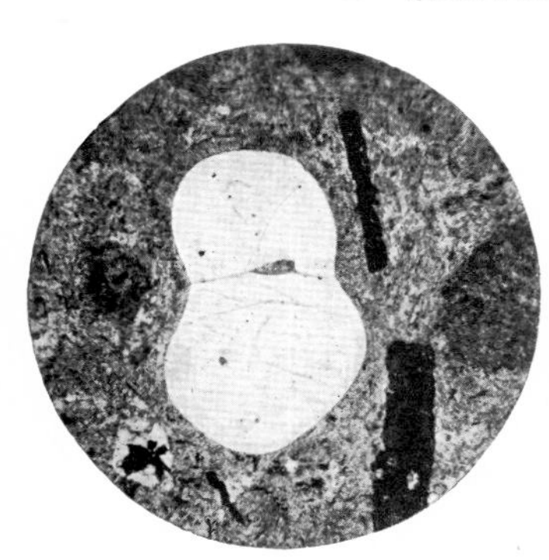

4. Quartz. Showing corrosion.

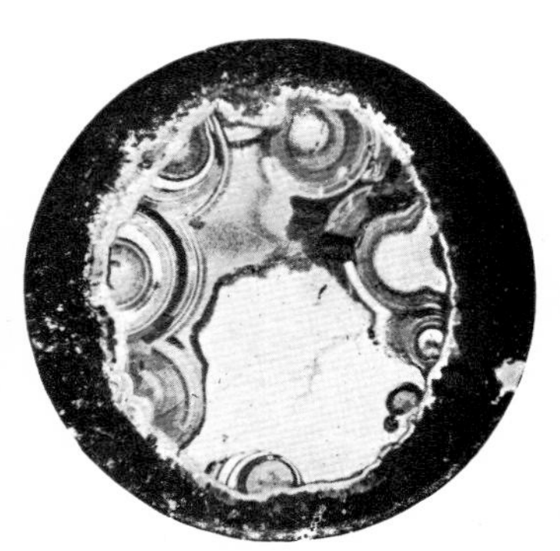

5. Stained chalcedony in amygdule.

PLATE VII

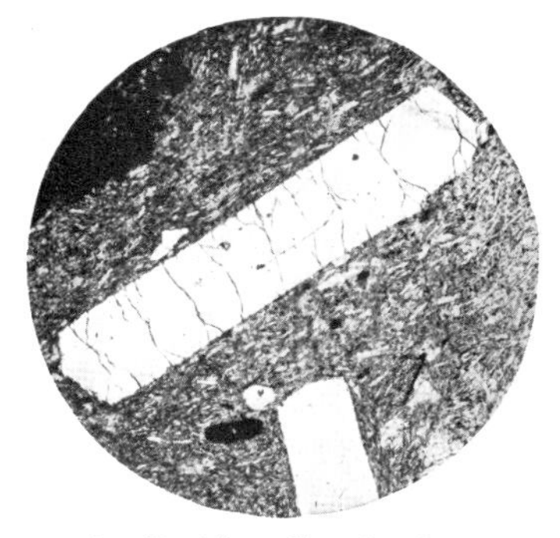

1. Sanidine. Showing form.

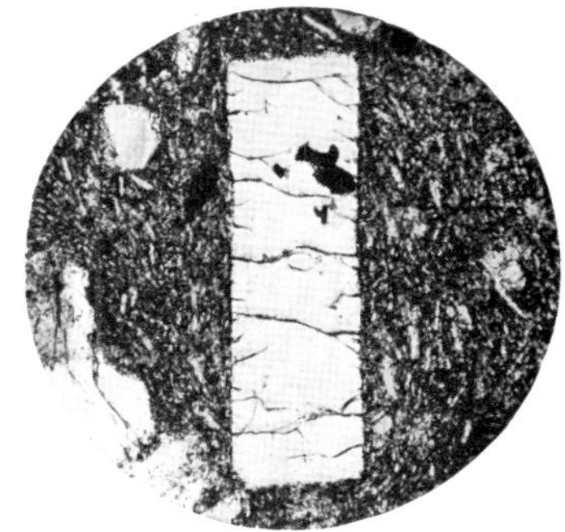

2. Sanidine. Showing form.

3. Sanidine. Crossed polarisers. Showing cleavages.

4. Orthoclase. Showing alteration and intergrowth with quartz.

5. Sanidine. Crossed polarisers. Showing Carlsbad twinning.

6. Sanidine. Crossed polarisers Showing Baveno twinning.

PLATE VIII

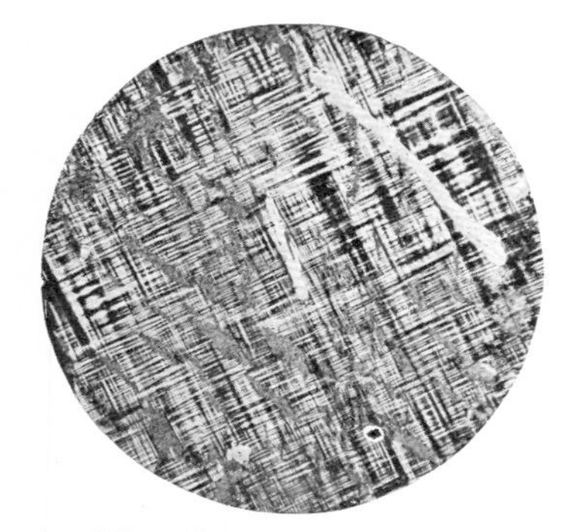

1. Microcline. Crossed polarisers. Showing cross-hatching.

2. Plagioclase. Crossed polarisers. Showing lamellar twinning.

3. Plagioclase. Showing zoning by alteration.

4. Plagioclase. Crossed polarisers. Showing Carlsbad-Albite twinning.

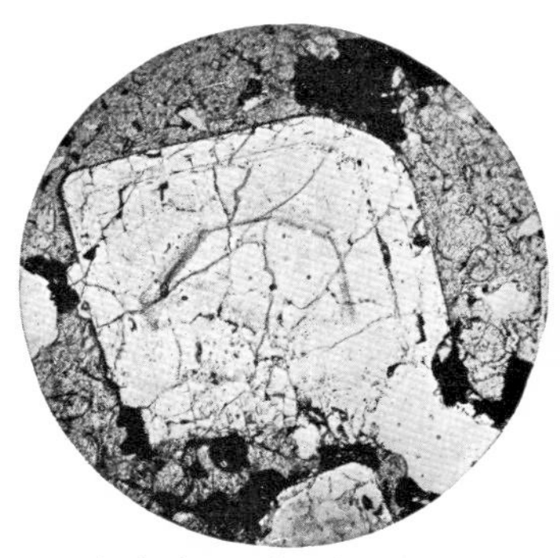

5. Plagioclase. Showing zoning by inclusions.

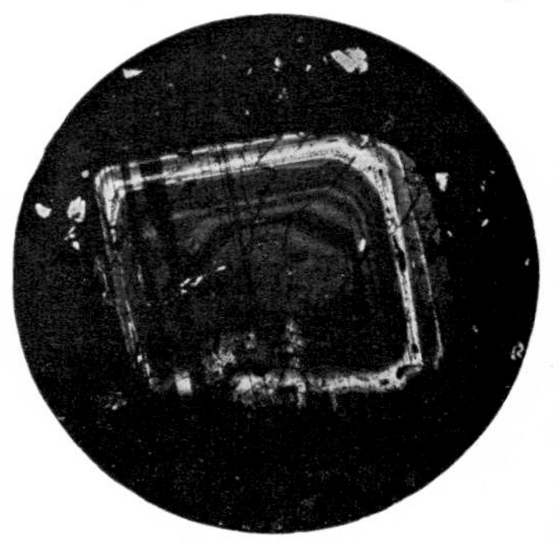

6. The same. Crossed polarisers. Showing zoning and lamellar twinning.

PLATE IX

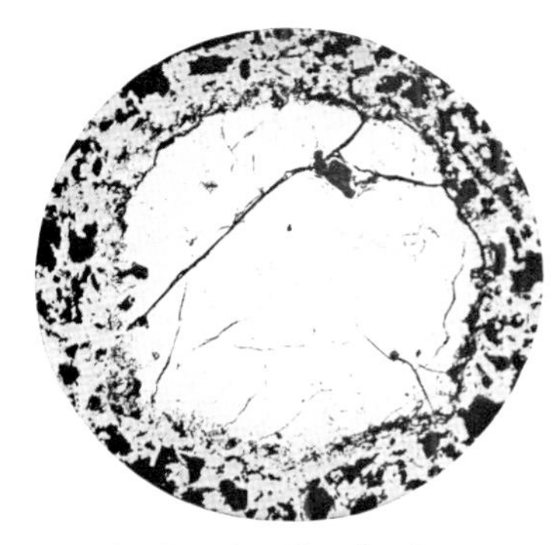

1. Leucite. Showing form.

2. The same. Crossed polarisers. Showing cross-hatching.

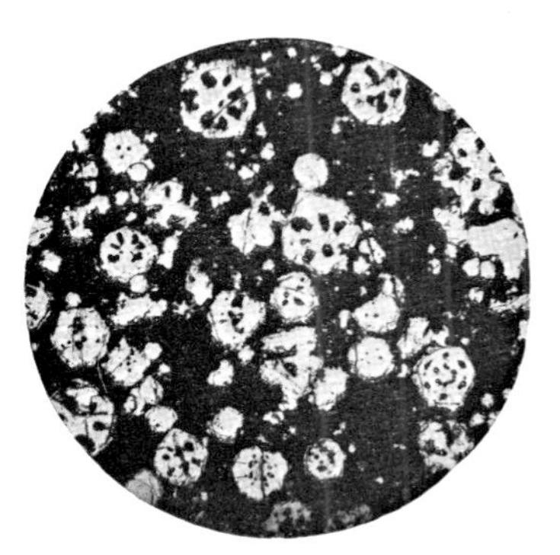

3. Leucite. Showing inclusions.

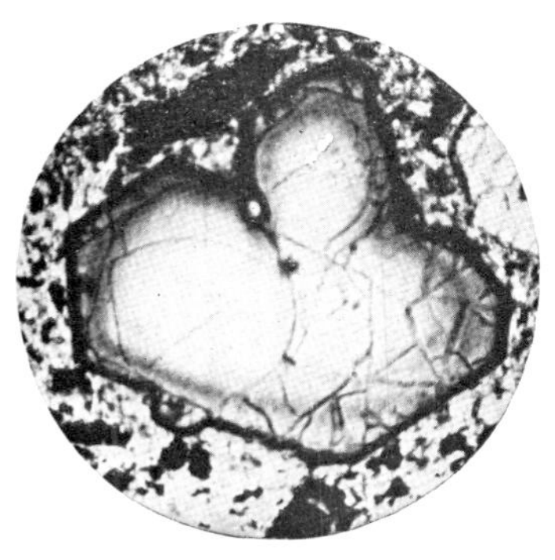

4. Nosean. Showing zoning.

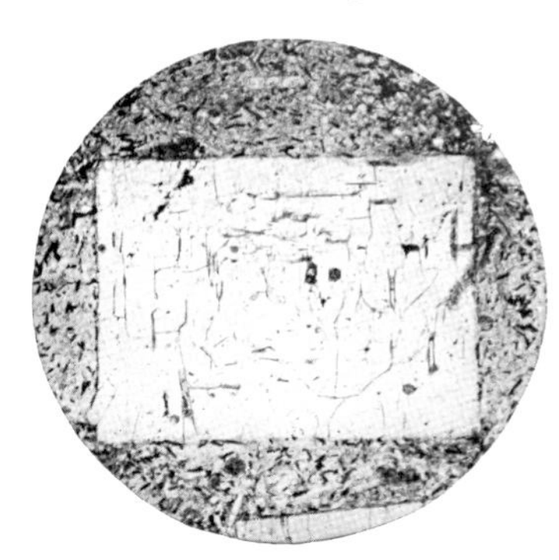

5. Nepheline. Showing form of vertical section.

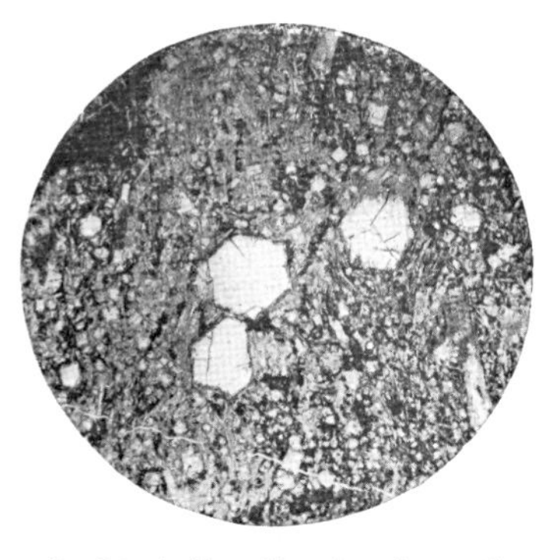

6. Nepheline. Showing form of basal sections.

PLATE X

1. Garnet (Melanite). Showing form and zoning.

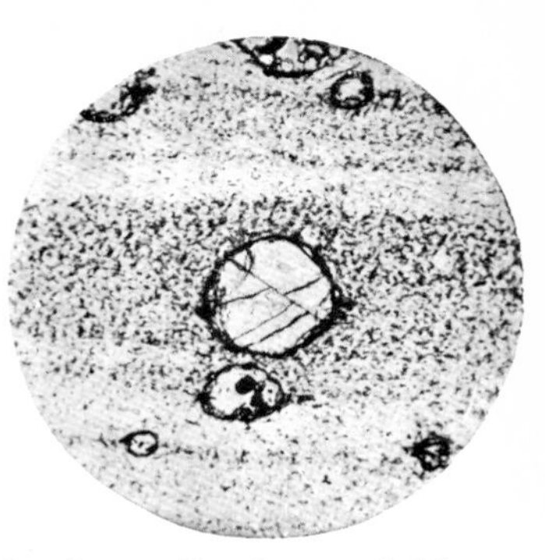

2. Garnet. Showing rounded form and high refractive index.

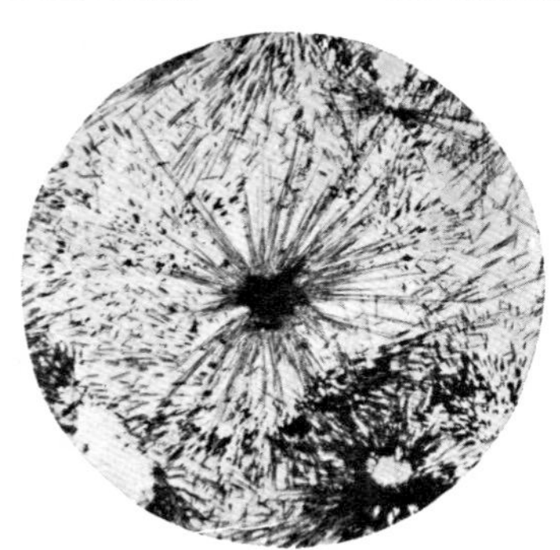

3. Tourmaline. Showing acicular and radiating habit.

4. Tourmaline, vertical section. Showing form.

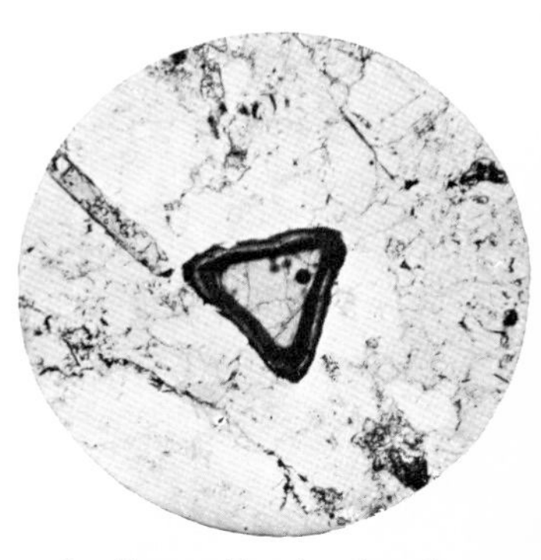

5. Tourmaline, basal section. Showing form and zoning.

PLATE XI

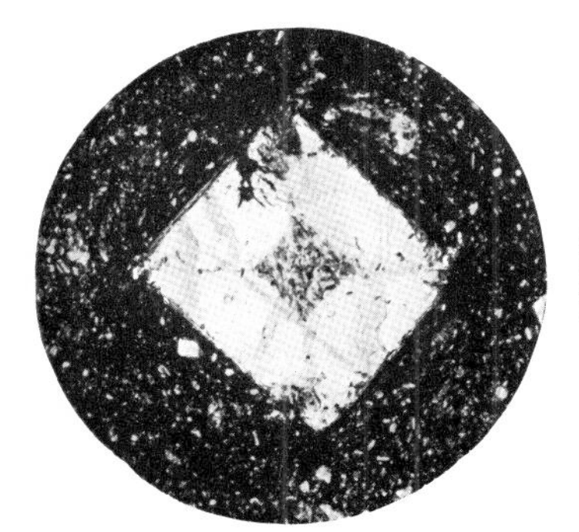

1. Chiastolite, basal section. Showing form and inclusions.

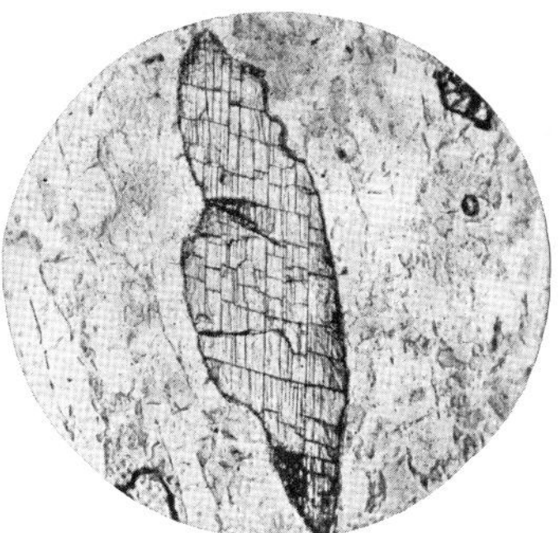

2. Kyanite. Showing cleavages and high refractive index.

3. Epidote. Showing form and cleavage.

4. Epidote. Showing occurrence in amygdule.

PLATE XII

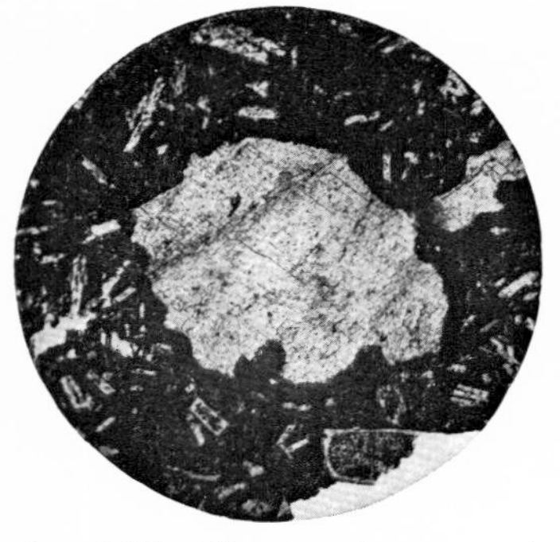

1. Calcite. Showing cleavages and the occurrence in amygdule.

2. Calcite. Crossed polarisers. Showing lamellar twinning.

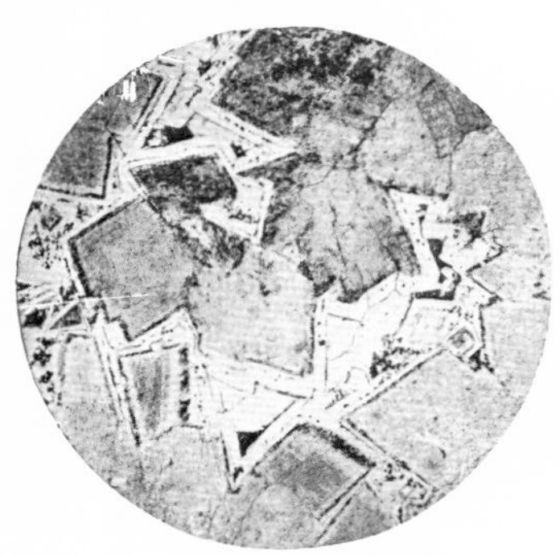

3. Dolomite. Showing form and zoning.

4. Zeolites. Showing occurrence with vermicular chlorite in amygdule.

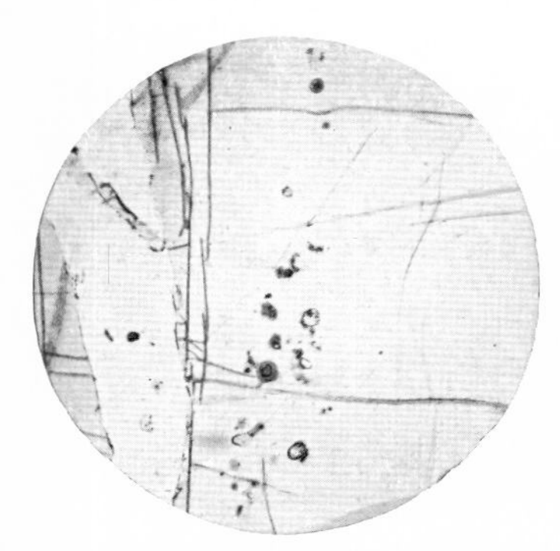

5. Cordierite. Showing haloes round inclusions.

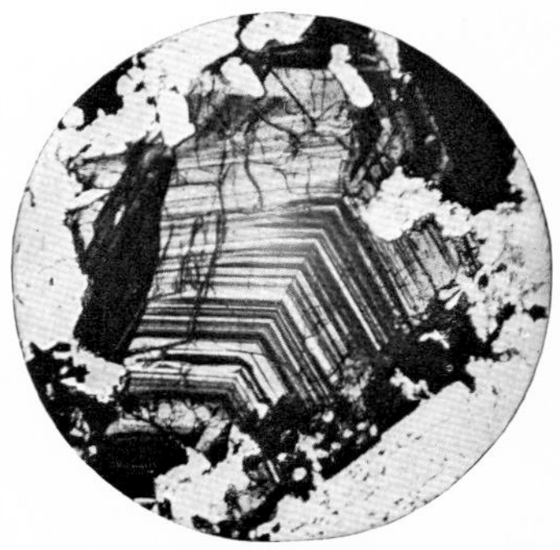

6. Cassiterite. Showing zoning.

extinguish, but yield a very distinctive deep blue interference colour, not found on the birefringence colour chart.

Besides occurring in calcareous schists in areas of regional metamorphism, zoisite is common in the more ancient basic igneous rocks, particularly those which have experienced a mild degree of metamorphism such as the Palæozoic basic sills and dykes in North Wales. In such occurrences zoisite figures prominently as an alteration product of basic felspar. It may pseudomorph the latter completely, or may be associated with secondary albite, calcite and other lime-bearing minerals in the aggregate formerly called saussurite.

Epidote, $(OH)Ca_2(Al,Fe)_3Si_3O_{12}$, is also characteristically prismatic in habit, with a very dark rich green colour and brilliant vitreous lustre. The crystals are usually deeply striated parallel to the length, while re-entrant notches at the ends of a crystal indicate twinning. The unusual feature is the fact that the crystals are *elongated parallel to the b-axis.* This corresponds to the 'Y' vibration direction, as the optic axial plane is parallel to the side-pinacoid. Consequently elongated sections in the zone parallel to the *b*-axis show straight extinction. Although often nearly black in hand-specimens, epidote is usually colourless in thin section, except in iron-rich varieties which may show a faint greenish yellow tint, and display slight pleochroism. This pleochroism is strikingly displayed by natural crystals which are not too thick, when held up to a strong light and rotated about the *b* (Y) axis. As the crystal is rotated, light corresponding successively to X and Z will reach the eye. These two vibrations are differently absorbed: for the one the colour transmitted is rich green; for the other, deep brown. This observation, which requires no apparatus, serves to differentiate between tourmaline and epidote, selected crystals of which may be closely similar in general appearance.

The optical properties of epidote vary with the content of iron. Iron-free epidote is called **clino-zoisite.** It is colourless in thin section, with strong relief, but weak to moderate birefringence with nα 1·717 and nγ 1·723. With increasing iron both relief and birefringence increase, the grains show a light yellow colour and are pleochroic. In ordinary epidote, the birefringence is strong, with nα 1·730 and nγ 1·770. Variations in the iron content within a single grain cause appreciable differences in interference colour: this is a useful diagnostic feature.

In igneous rocks of widely different composition epidote occurs as an alteration product of the common ferro-magnesian minerals, augite and hornblende, and represents the original CaAl content. The magnesian portion of the original augite or hornblende is now represented by chlorite, in which the epidote is typically embedded in the form of scattered grains or perhaps more typically, stellate aggregates of small prisms. If the original mineral contained manganese, the epidote is associated with a much rarer reddish brown or brown member of the group, a manganese epidote, withamite.

THE OPAQUE MINERALS

The detailed study of opaque minerals is a specialised branch of microscopy, demanding the use of a modified petrological microscope, using reflected polarised light. It is therefore outside the scope of this book, though the metallurgical microscope opens a new and valuable field of study.

For our present purposes it will be sufficient to indicate ways and means of identifying the widely distributed **iron ores**, one or other of which is seldom absent from a rock-section. The mineral species involved are magnetite, ilmenite, pyrite, hæmatite and limonite. All appear black by transmitted light.

Magnetite, Fe_3O_4, Cubic, holosymmetric. This oxide of iron is a member of the **spinel** group of minerals, with the general formula $R^{++}OR^{+++}{}_2O_3$, in which the divalent R^{++} is commonly Mg or Fe, less commonly Mn or Zn.

The trivalent R^{+++} is Al, Fe or Cr in different cases. The best-known spinel in mineral collections occurs as beautifully coloured small red transparent octahedra, sometimes with the edges modified by the rhomdodecahedron, and sometimes twinned. The spinels encountered in thin rock sections, however, tend to be very strongly coloured and at least semi-opaque: indeed, they often transmit light only on the thin edges. Typical colours encountered include brownish black (in chromite), very dark green and plum-purple.

Occurrence. Spinels occur as occasional accessory minerals in very basic igneous rocks; but are more widely distributed as products of thermal metamorphism of sediments of the appropriate composition. Dark green spinel frequently accompanies cordierite and occurs in the form of swarms of small octahedra embedded in the latter. Magnetite itself shows the same crystallographic characters, but is much more common. It may be seen at its best perhaps in basic lavas: in the fine-grained groundmass of basalt small triangular, nearly square and six-sided sections of magnetite octahedra are usually plentifully scattered throughout. **Chromite,** $FeO.Cr_2O_3$, in thin section may be quite opaque like magnetite, and therefore indistinguishable from the latter by ordinary optical tests. The only way to be quite certain is to isolate the mineral and test it for magnetism with a bar magnet, or chemically for chromium. It has been established, however, that the small opaque octahedra occurring with olivine in the almost monomineralic olivinite ('dunite') are chromite. Another olivine-magnetite association may be referred to. When olivine is converted into serpentine, the iron originally in combination is not accommodated in the serpentine, but appears as newly formed minute crystals of magnetite embedded in it.

Ilmenite (titaniferous iron-ore), essentially $FeO.TiO_2$, but containing a variable amount of Fe_2O_3 in addition. Trigonal. Ilmenite, when quite fresh, may be indistinguishable from magnetite by optical tests, and a microchemical test is essential. If altered, however, it is easily and surely identified by the presence of **leucoxene**—an opaque whitish encrustation. As the alteration progresses, the originally opaque mineral becomes translucent and brownish in colour. The change is due to the separation of the titanium from the iron: the former ends up as sphene, the latter as magnetite. Again, although fresh ilmenite gives no indication under the microscope of its crystal symmetry it is often clearly apparent in the altered material. Ilmenite occurs typically as an accessory in deep-seated basic igneous rocks, and its presence may be suspected if, in the section, other Ti-bearing minerals such as titanaugite and barkevikite have been identified. Without the guide of associated Ti-bearing silicates or of leucoxene, ilmenite is difficult to distinguish from magnetite particularly when both minerals are together in the same rock, as they frequently are in gabbros and anorthosites.

Pyrite, Cubic, FeS_2. In igneous rocks pyrite is invariably secondary and is therefore restricted to altered rocks. It is quite opaque in thin section, but is readily identified by its light brassy appearance under reflected light.

Pyrite may be a prominent constituent of heavy-mineral suites and in these circumstances the observer has the additional evidence of the three-dimensional form of the crystals. Small striated cubes, often with the edges modified by other forms, and identical in all but size with the well-known pyrites in mineral collections, are very distinctive among the opaque minerals.

Hæmatite, Fe_2O_3, Trigonal. Beautifully crystalline specimens of hæmatite are well-known objects in mineral collections, but it is disappointing material in rock-sections, usually occurring only as minute grains or disseminated specks filling the role of pigment. Thus the bright red streaks and blotches in the Cornish serpentines consist of hæmatite. In these conditions the hæmatite is bright red in thin section; but when more massive it may be black and opaque.

Limonite is the name commonly applied to the hydrous iron-oxides occurring in rocks of all kinds. In this sense the name also covers goethite, an Orthorhombic mineral with the composition $Fe_2O_3.H_2O$. This is pleochroic in thin slice, from yellow to brown; while limonite itself, with a higher content of water, is apparently isotropic, and without reaction on polarised light.

Summarily, the opaque iron ores may be difficult to identify in any given case; but usually there is some feature —it may be only that of association with other minerals— which suggests a diagnosis. In difficult cases it is a good plan to tilt the microscope so that the slide is illuminated with strong incident light, when the surface features may be observed. Under these conditions, magnetite appears steely grey-black; ilmenite may appear purplish or, if altered, whitish through the development of leucoxene; pyrite is brassy, hæmatite *may* be red and limonite, including goethite, yellow to brown.

THE ROCK-FORMING CARBONATES

Natural carbonates of calcium, magnesium and iron are the chief carbonates which are important as rock-builders. Of these, calcite, $CaCO_3$ is dominant; but locally siderite (chalybite), the corresponding iron carbonate, and dolomite, the double carbonate of calcium and magnesium, are important. These minerals are distinctive in natural crystals or aggregates, by reason of differences in crystal characters and in colour; but under the microscope distinction between them is not easy, and may involve the accurate measurement of the refractive indices, using immersion methods.

Calcite, $CaCO_3$, Trigonal (Pl. XII). Calcite is one of the commonest minerals and one of the best known. It is the principal constituent of limestones and their metamorphic equivalents, and is also widely distributed in veins. In igneous rocks it is rarely a primary constituent, though it is abundant in the intrusive carbonate-rocks, of doubtful origin, termed carbonatites. Despite the great diversity of beautiful crystals found in vugs and veins and figuring prominently in mineral collections, calcite occurring in rock sections hardly ever displays crystal form. But although the external shape is wanting, the internal structure is indicated by a ubiquitous, perfect rhombohedral cleavage, which is always conspicuous. It may be noted that even in plane-polarised light these cleavages often show interference colours—a thin-film effect.

The chief optical properties have already been discussed, so that only a brief summary is necessary here. Calcite is colourless, uniaxial negative, and because of the unusually wide separation of the extreme refractive indices, grains show a strikingly different surface relief as the stage is rotated. In one position the grain may be practically invisible, but in another the relief is strong, and the grain boundary and cleavage traces stand out very clearly. Multiple twinning is common. These features, together with the high order light pink, green and greyish white interference colours, make calcite one of the easiest minerals to identify.

Dolomite, $CaCO_3.MgCO_3$ — the double carbonate of calcium and magnesium. Trigonal. By contrast with the great diversity of crystal form displayed by calcite, dolomite develops only one form—the unit rhombohedron (that to which the cleavages are parallel). Again unlike calcite, dolomite in thin sections normally occurs in the form of small perfect unit rhombohedra (Pl. XII), usually embedded in a matrix of powdery calcite which it has replaced. It can also often be observed to have replaced the original substance (often aragonite) of fossil shells. Apart from this difference in crystal habit, dolomite is closely similar to calcite: indeed the two are difficult to distinguish without the aid of a staining or microchemical test.

Siderite, $FeCO_3$, Trigonal, isomorphous with dolomite, and showing precisely the same crystal form — the unit (cleavage) rhombohedron. A tendency to alteration, at least superficially, into limonite is a useful aid in identifying this carbonate, as it colours the crystals brown, in some cases a deep chocolate brown. In thin section also, alteration into limonite is the best indication that a given carbonate is siderite. A valuable confirmatory observation concerns the refractive indices, which are considerably above those of slide-mounting media: n_ω 1·875; n_ϵ 1·633. This applies, it should be noted, to *both* indices — compare calcite. Further, the birefringence is extremely high, 0·2; while unlike dolomite, lamellar twinning is common.

SUGGESTED REFERENCES

F. D. Bloss. An introduction to the methods of optical crystallography. Holt, Rinehart and Winston. 1961.

E. S. Dana and W. E. Ford. A textbook of mineralogy. John Wiley & Sons Inc. Fourth Edition. 1932.

A. F. Hallimond. Manual of the polarising microscope. Cooke, Troughton and Simms, Ltd. Second edition. Reprinted. 1956.

N. H. Hartshorn and A. Stuart. Crystals and the polarizing microscope. Edward Arnold & Co. Second edition. 1950.

F. H. Hatch, A. K. Wells and M. K. Wells. Petrology of the Igneous Rocks. Thomas Murby and Co. Twelfth edition. 1961.

H. H. Read. Rutley's Elements of Mineralogy. Thomas Murby & Co. Twenty-fourth edition. 1948.

P. F. Kerr. Optical mineralogy. McGraw-Hill Book Co. Inc. Third edition. 1959.

A. N. Winchell and H. Winchell. Elements of optical mineralogy. Part II. Descriptions of Minerals. John Wiley & Sons Inc. Fourth edition. 1951.

INDEX

GEORGE ALLEN & UNWIN LTD

Head office:
40 Museum Street, London, W.C.1
Telephone: 01-405 8577

Sales, Distribution and Accounts Departments
Park Lane, Hemel Hempstead, Herts.
Telephone: 0442 2344

Athens: 7 Stadiou Street, Athens 125
Auckland: P.O. Box 36013, Northcote, Auckland 9
Barbados: P.O. Box 222, Bridgetown
Beirut: Deeb Building, Jeanne d'Arc Street
Bombay: 103/5 Fort Street, Bombay 1
Calcutta: 285J Bepin Behari Ganguli Street, Calcutta 12
P.O. Box 23134, Joubert Park, Johannesburg, South Africa
Dacca: Alico Building, 18 Montijheel, Dacca 2
Delhi: 1/18 B Asaf Ali Road, New Delhi 1
Ibadan: P.O. Box 62
Karachi: Karachi Chambers, McLeod Road
Lahore: 22 Falettis' Hotel, Egerton Road
Madras 2/18 Mount Road, Madras 2
Manila: P.O. Box 157, Quezon City D-502
Mexico: Libreria Britanica, S.A., Separio Rendon 125, Mexico 4, D.F.
Nairobi: P.O. Box 30583
Ontario: 2330 Midland Avenue, Agincourt
Rio de Janeiro: Caixa Postal 2537-Zc-oo
Singapore: 248c–6 Orchard Road, Singapore
Sydney, N.S.W.: Bradbury House, 55 York Street
Tokyo: C.P.O. Box 1728, Tokyo 100-91